AF769670

Studies in Medieval History and Culture

Edited by

Francis G. Gentry
Pennsylvania State University

A Routledge Series

STUDIES IN MEDIEVAL HISTORY AND CULTURE

FRANCIS G. GENTRY, General Editor

THE REPRODUCTIVE UNCONSCIOUS IN LATE MEDIEVAL AND EARLY MODERN ENGLAND
Jennifer Wynne Hellwarth

MYSTICAL LANGUAGE OF SENSE IN THE LATER MIDDLE AGES
Gordon Rudy

FAIR AND VARIED FORMS
Visual Textuality in Medieval Illustrated Manuscripts
Mary C. Olson

THE CONTESTED THEOLOGICAL AUTHORITY OF THOMAS AQUINAS
The Controversies between Hervaeus Natalis and Durandis of St. Pourçain
Elizabeth Lowe

BODY AND SACRED PLACE IN MEDIEVAL EUROPE, 1100-1389
Dawn Marie Hayes

WOMEN OF THE HUMILIATI
A Lay Religious Order in Medieval Civic Life
Sally Mayall Brasher

CONSUMING PASSIONS
The Uses of Cannibalism in Late Medieval and Early Modern Europe
Merrall Llewelyn Price

LITERARY HYBRIDS
Crossdressing, Shapeshifting, and Indeterminacy in Medieval and Modern French Narrative
Erika E. Hess

THE KING'S TWO MAPS
Cartography and Culture in Thirteenth-Century England
Daniel Birkholz

PESTILENCE IN MEDIEVAL AND EARLY MODERN ENGLISH LITERATURE
Bryon Lee Grigsby

RACE AND ETHNICITY IN ANGLO-SAXON LITERATURE
Stephen J. Harris

ASPECTS OF LOVE IN JOHN GOWER'S CONFESSIO AMANTIS
Ellen Shaw Bakalian

THE MEDIEVAL TRADITION OF THEBES
History and Narrative in the OF Roman de Thèbes, Boccaccio, Chaucer, and Lydgate
Dominique Battles

WORLDS MADE FLESH
Reading Medieval Manuscript Culture
Lauryn S. Mayer

EMPOWERING COLLABORATIONS
Writing Partnerships between Religious Women and Scribes in the Middle Ages
Kimberly M. Benedict

THE EPISTEMOLOGY OF THE MONSTROUS IN THE MIDDLE AGES
Lisa Verner

THE WATER SUPPLY SYSTEM OF SIENA, ITALY
The Medieval Roots of the Modern Networked City
Michael P. Kucher

The Water Supply System of Siena, Italy

The Medieval Roots of the Modern Networked City

Michael P. Kucher

Routledge
New York & London

Published in 2005 by
Routledge
270 Madison Avenue
New York, NY 10016
www.routledge-ny.com

Published in Great Britain by
Routledge
2 Park Square
Milton Park, Abingdon
Oxon OX14 4RN
www.routledge.co.uk

Copyright © 2005 by Taylor & Francis Group, a Division of T&F Informa.
Routledge is an imprint of the Taylor & Francis Group.

Printed in the United States of America on acid-free paper.

All rights reserved. No part of this book may be reprinted or reproduced or utilized in any form or by any electronic, mechanical, or other means, now known or hereafter invented, including photocopying and recording, or in any information storage or retrieval system, without permission in writing from the publishers.

10 9 8 7 6 5 4 3 2 1

Library of Congress Cataloging-in-Publication Data

Kucher, Michael P.
The water supply system of Siena, Italy : the medieval roots of the modern networked city / by Michael P. Kucher.
p. cm. -- (Studies in medieval history and culture ; v. 29)
ISBN 0-415-97166-7 (alk. paper) 1. Water-supply--Italy--Siena--History--To 1500. 2. Fountains--Italy--Siena--History. 3. Aqueducts--Italy--Siena--History. 4. Siena (Italy)--History. I. Title. II. Series.

TD280.S54K83 2005
628.1'4'094558--dc22 2005018359

Contents

List of Figures

Series Editor's Foreword

Far from providing just a musty whiff of yesteryear, research in Medieval Studies enters the new century as fresh and vigorous as never before. Scholars representing all disciplines and generations are consistently producing works of research of the highest caliber, utilizing new approaches and methodologies. Volumes in the Medieval History and Culture series include studies on individual works and authors of Latin and vernacular literatures, historical personalities and events, theological and philosophical issues, and new critical approaches to medieval literature and culture.

Momentous changes have occurred in Medieval Studies in the past thirty years, in teaching as well as in scholarship. The Medieval History and Culture series enhances research in the field by providing an outlet for monographs by scholars in the early stages of their careers on all topics related to the broad scope of Medieval Studies, while at the same time pointing to and highlighting new directions that will shape and define scholarly discourse in the future.

Francis G. Gentry

Preface and Acknowledgments

The United Nations declared 2003 the "International Year of Freshwater." It was only fitting that I began the process of revising this manuscript for publication in that year. The residents of medieval Siena found a brilliant, and for its day, "high-tech" system of water supply that had minimal impacts on the environment. If a millennium of continuous operation is evidence of a sustainable system, then one could describe Siena's system with that increasingly popular term. One need only contrast it with the worrisome trend of "mining" glacial deposits of water around the world. Edward Goldsmith described the lessons such systems could teach us today in his aptly titled, "Learning to Live with Nature." There is much we can learn from Siena.

This book grew out of my dissertation in the department of history at the University of Delaware, which in turn emerged as the inevitable result of having caught a glimpse of the City of the Virgin in the summer of 1987. I would not have made that trip except for listening to advice from Oscar Shoenfeld that if I was going to Italy, I must go to Siena. Little did I know at the time it would be the first of many visits and the beginning of a long fascination with that city and the processes that underlay its design and construction.

Soon after beginning the research for this project it became clear that although many of the sources I would need to consult had been published, they remained scattered in hard to find volumes. Most of the unpublished sources, however, remain in Siena, where they are very well cared for by the dedicated archivists and librarians of that city. The secondary literature used for this work spans several disciplines and at least two continents. Geographic, disciplinary, and linguistic barriers have slowed the synthesis of this knowledge. It is my hope that this work will provide a starting point for that synthesis and integration of scholarship from disparate disciplines. I hope I have succeeded in answering some of the questions about the social context and fecundity of Siena's artist-engineers.

It is humbling to recall the number of debts I have incurred in even such a modest project as this book and it gives me great pleasure to thank some of the people and institutions who have helped me along the way: the mentors and editors who guided me; friends, family, and colleagues on two continents who provided encouragement; the librarians and archivists who went out of their way to make various books, articles, maps, and manuscripts available; the institutions who support scholarship; and the citizens of Siena who made me feel welcome and who had the patience to explain some of that incredible city's subtleties to an outsider. Needless to say, almost no one fit neatly into one of the above groups. To describe the ways in which each of the following assisted me would require a chapter-length treatment. I can do little more than list the names of the many who have so generously assisted me and dread that I have forgotten others.

Many institutions made this work possible, especially the National Science Foundation (award #9222747), National Endowment for the Humanities, Hagley Fellowship, Fulbright-Hays Commission, Bassett Ferguson Fund, History Department at the University of Delaware, New York Public Library, Dibner Library and Special Collections of the Smithsonian Institution, University of Washington Rome Center, the Chancellor's office at the University of Washington, Tacoma, which provided a research leave to revise the manuscript for publication, the members of the Society for the History of Technology, anonymous referees of the *Journal of Urban History,* and of *Technology and Culture,* and the Monte dei Paschi di Siena.

Colleagues, teachers, mentors, editors, and friends have all made this work a stronger piece of scholarship. They include: Sara Adler, David Allmendinger, Hope Alswang, Max van Balgooy, George Basalla, Silvio Bedini, Luisa Bellaviti, Keith Benson, Renie Bierman, Elise Bowditch, William Bowsky, Jill Caskey, Brian Coffey, Liz Coleman, Jeffrey Collins, Trina Deines, Yonn Dierwechter, Larry Duggan, Gene Ferguson, Michael Forman, Larry Fuchsberg, Reed Geiger, Lisa Gitelman, David Goldfield, Carolyn Goldstein, George Gorse, Harvey Green, Cheryl Greengrove, Daryl Hafter, Tom Hankins, Amy Hardin, Stett Harrison, David Hounshell, Alice Hudson, Tom Hughes, Bill Leslie, Karl Hufbauer, Beth Kalikoff, Jim Karr, Roger Kennedy, Liz Kogen, Faith and Vladimir Kucher, Tim LeCain, Jean LeCorbeiller, Mark Levinson, Pam Long, Steve Lubar, Robert MacKay, Roberta Magnusson, Paul Mankiewicz, Mina Marefat, Michelle McFadden Clay McShane, María Rosa Menocal, John Muendel, Art Molella, Lynn Nelson, Ruth Oldenziel, Keiji Oka, Patricia Orendorf, John Peterson, Glen Porter, Bill Richardson, Heddy Richter, Katherine Rinne, Richard Rodgers, Gustina Scaglia, David Secord, Kim Sexton, Peggy Shaffer, Joe Shoenfeld,

John Staudenmaier, Randy Staudinger, Kathy Steen, Woody Sullivan, Lynn Untz, Kelly DeVries, Keith Ward, Susan Williams, Steve Winch, and Elizabeth Wyckoff.

Working close to home in well equipped university libraries is hard enough. Fitting one's office into forty kg of checked baggage presents its own challenges, ones that I would have never overcome without help from dozens of scholars, archivists, librarians, citizens, and fellow scholars in Italy. They include: Sonia Adorni Fineschi, Duccio Balestracci, Didier Boisseuil, Mario Brogi, Maria Assunta Ceppari, Roberta Bargagli, Donatella Ciampoli, Giulio Catoni, Giuseppe Chironi, Claudia Fanciulli, Riccardo Giacopelli, Bernardo Giorgi, Fratelli Losi, Luca Luchini, the Mayor's Office of Siena, Stefano Moscadelli, Ludwin Paardekooper, Roberta Panzanelli, Raffaella Pulejo, Francesco, Maura, and Luca Rinaldi, members of the Associazione La Diana, Franco Runfola, Nuovafotografia, Fulvia Sussi, Patrizia Turrini, Adele and Marinella Vaglio, and Carla Zarrilli.

I benefitted from admission to many incredible libraries, including those of UCLA, Pomona College, University of Delaware, the University of Washington, University of Nevada, and the Huntington Library, National Library of Medicine, Newberry Library, Hagley Museum and Library, the Smithsonian Libraries, the Library of Congress, Huntington Library, the National Library of Medicine, the Division of Prints and Photographs and the Maps Division at the New York Public Library. Without skilled librarians, I would have never found what I was looking for. Thanks to Bridget Burke, Stephen Greenberg, Nancy Gwinn, Suzanne Klinger, Matt Knutzen, David Langenberg, Charles Lord, Nancy Matthews, Theresa Mudrock, Heddy Richter, Scott Spooner, Rhoda Ratner, Jim Roan, Ellen Wells, and Katie Whitson.

Without intending offence to any of the many wonderful archives that have so graciously allowed me to be a guest (and at the risk of repeating the observations of other historians), I need to say that there is no finer archive in the world in which to do research than the Archivio di Stato in Siena and no better city in the world for a foreigner to arrive seeking a temporary home and community. Siena's motto, *Cor magis tibi Sena pandit* is not a promise or aspiration, it is a simple statement of fact.

Without a publisher, most scholarship goes unnoticed. Thanks to Paul Johnson, Frank Gentry, Kimberly Guinta, and William Germano. I was lucky enough for this work to find a wonderful press in Routledge.

My greatest debt is to the world's most patient woman, my wife Julie.

Omissions and errors are my own.

Chapter One

Introduction

"In the fifteenth century, the technology of Italy was even richer than that of Germany, yet its records are little known. The lavishly illustrated notebook of Giovanni Fontana preserved in Munich (Staatsbibliothek, Codex icon. 242) has been scantily examined. From the 1440s until the 1480s, Siena, which was best represented by Mariano di Jacopo detto il Taccola and by Francesco di Giorgio, seems to have been the chief source of innovation. . . . Their impact upon the general culture of Europe was as great as that of the contemporary Italian humanists, artists, and musicians, yet their history remains completely unwritten." (Lynn White, Jr., "The Flavor of Early-Renaissance Technology")

Lynn White, Jr. made these suggestions at a time when the notebooks of Villard de Honnecourt, the *Bellifortis* of Konrad Kyeser, and the notebook of the anonymous German engineer of the Hussite Wars were just beginning to receive the attention they deserved. White was trying to call his readers' attention to the impact of engineers, such as Taccola and Francesco di Giorgio Martini, south of the Alps. He argued that our lack of a history of Italian engineering formed "the greatest single lacuna in our understanding of the diffusion of the Italian Renaissance." In order to explain why Siena—a quiet hill town in Tuscany, best known today for the remarkable state of preservation of its historic core—might have been the "chief source of innovation" in technology during the fifteenth century, it is necessary to discuss the city's history leading up to this period of intellectual fecundity. The surviving remains of its medieval built environment, including palaces, the cathedral, roads, walls, fountains, and *piazze,* provide the first clues that Siena was once a thriving political, commercial, and cultural center that vied with Florence for control of southern Tuscany. With foundations going back to at least the early Middle Ages, Siena enlarged its walls frequently, attesting to its demographic and geographic growth. The city reached the height of its power and influence between 1250 and 1350. During that period the city controlled the region between the Tyrrhenian coast to the west and the Apennines to the east, Florence's territory to the north and Arezzo's to the south. Its economy was based on agriculture, government, religion, education, banking, some manufacturing, and commerce. The bishop, hospital, university, and numerous religious foundations did much to shape its culture, but the government—which jealously guarded its independence from the church, the Roman Empire, and the influences of local magnates—remained the single most powerful force in shaping Sienese society and culture. The plague of 1348 brought with it not only demographic collapse and the end of a long period of political stability; it also signaled the beginning of Siena's decline as a city-state, a process that would be brought to a close with its loss of independence, to Florence, in 1555. What was it about Siena that made it in some way especially suited to become a center of innovation?[1]

Two interpretations of Siena's place in Italian history prevail today: Siena as a unique manifestation of late-medieval urban culture, second only to Florence, and Siena as a typical Tuscan hill town. Unfortunately, neither of them does much to elucidate the issues Lynn White raised in 1972. The first, long held by native Sienese seeking to recover Siena at the peak of its power and influence, emphasizes a romantic, idealized medieval city, unified by religion and ruled by concord. Carlo Falletti-Fossati's imaginative 1881 reconstruction of urban life in medieval Siena is the best example of the romantic perspective. The second image of Siena, promulgated in the writings of British and American historians, is that of Siena as just another Tuscan hill town, albeit with regard to cultural achievements second only to Florence, the city that for most people epitomizes the achievements of late-medieval Tuscany. Historians such as R. Langton Douglas and Ferdinand Schevill, writing at the beginning of this century, set the tone for this balanced approach that saw nothing absolutely unique about Siena.[2]

A study of Siena's water supply system offers one way to determine the city's uniqueness or sameness and to test White's theory that Siena was a major source of technical innovation in the fifteenth century. Contrary to the long-held assumption among its users and early historians that the Sienese water supply is exceptional, this study finds the Sienese system to be typical of those hill towns in central Italy that share Siena's geological features. However, rather than a single homogeneous system, as most historians have portrayed it, Siena derived its water from a combination of wells, cisterns, springs, and aqueducts. The significance of Siena's water supply system lies in the scale of the system, the ways in which it challenged the engineering skills of its creators, and the new environment of municipal technical and artistic patronage that it and other urban projects fostered.[3]

One cannot examine Siena's water supply system without some basic background information regarding Siena's social, cultural, and economic history. Siena's early-medieval economy had long rested on several factors, chief among them a combination of agriculture, the maintenance of a local market, and its function as the administrative center for the diocese. In the eleventh century, the city's location along the main route from northern Europe to Rome allowed for the diversification of its economy into financial and commercial activities and the expansion of some of its craft-based industries, especially textile production. However, its hilltop location prevented it from catching the next wave of economic development—the application of water power to manufacturing processes—because the best mill sites in its territory were several kilometers beyond the city's walls. Nonetheless, the combination of trade, agriculture, banking, textile production, construction, and numerous religious foundations allowed the city to grow to nearly fifty thousand inhabitants by the 1320s—only half that of Florence—a figure that was not attained again until the twentieth century.[4]

The technology that provided the material means for this expansion was, to a greater extent than has been previously acknowledged, foreign. As the scale of cities grew, their needs began to resemble more closely those of Levantine, Asian, and Iberian cities. Siena, positioned on one of the most important roads in the Middle Ages, was an active trader and financier in the expanding "global" economy, roles that put it in touch with not only towns along that road, like its termini Paris and Rome, but also with trading partners outside of Christian Europe, such as cities on the Iberian Peninsula, Sicily, and the Levant.

Siena's role as a commercial center brought it into contact with cities on the peninsula like Pisa and Genoa as well as foreign ports. Siena bought grain from Sicily, alum from Egypt and the near East, and luxury goods from Asia. Siena imported engineering and artistic talent from other cities, such as Niccolò and Giovanni Pisano and Gentile da Fabriano from their eponymous homes and Donatello, Lorenzo Ghiberti, and Leon Battista Alberti from Florence. It also loaned its most talented engineers to other cities, such as Jacopo della Quercia to Lucca, Francesco di Giorgio to Urbino, Lando di Pietro to Naples, and Lorenzo Maitani to Orvieto. The city would recall its artists from abroad when it needed them.[5]

Despite flourishing trade relations during the Middle Ages, the brief period during which Siena was able to sustain its peak population of fifty thousand, along with the centuries it took for this number to rebound, suggests

Figure 1.1 Siena and Region. Siena straddles the watersheds of the Arno and Ombrone. From Schevill, *Siena.*

that Siena had approached the physical limits of its site and exceeded its territory's "carrying capacity," reaching what David Herlihy called a point of demographic deadlock.[6]

The facts of Siena's foundation and origins remain unclear. Although believed by some Sienese to have been founded by the Etruscans and colonized by the Romans, in absence of firm evidence, the scholarly consensus as to Siena's origins is vague, pointing to Late Antiquity or the early Middle Ages. For centuries Siena functioned as a local administrative center for the Church and the seat of the bishop, whose office was evidently first established in the fourth century. The earliest physical remains of the city date from perhaps a century later. Siena's political evolution parallels that of many Italian city-states. Despite nominal jurisdiction of the Roman emperors, in practical terms both civil and ecclesiastical power rested with the bishop until the middle of the twelfth century. Economic growth and urbanization created new wealth and began to shift power to people outside the ecclesiastical administration.[7]

Frederick Barbarossa's twelfth-century concession of self-government was one of several steps in a process of creating strong secular bodies of government already well underway. The ultimate result of more than a century of conflict over who would exercise secular power was a dynamic equilibrium between church and state manifested in frequent collaboration between civil and ecclesiastical authorities. Parity between the two poles of power more or less prevailed until Siena's submission to Florence in 1555. The first clue that the bishop of Siena had begun to share his power came in 1124 when a number of *boni homines* (good men) accompanied bishop Gualfredus to Rome. By 1156 powerful citizens began to function as advisors, or consuls, to the bishop. By 1167 the consuls ousted the bishop (who died in exile in 1170) and firmly established secular control of the city, forming themselves into the secular Council of the Bell, also known as the General Council, the "chief deliberative and legislative body." By 1179 Siena had a constitution and a seal. The contest between the bishop and emerging civil authorities was repeated in numerous Tuscan cities in the eleventh and twelfth centuries. The Emperor legitimatized this secular seizure of power in Siena with a charter in 1186. The bishop soon returned to his see, but he never wielded secular power again, and the city played a considerable role in decisions that used to be entirely his, such as the design and construction of the new cathedral or the collection of revenues from his fiefdom, the Vescovado.[8]

By 1199, Siena had joined many of its neighboring city-states in hiring a *Podestà,* a sort of city manager who was frequently an outsider, to carry out the day-to-day administration and judicial functions of city government. By 1211, Siena required this appointed official to be a foreigner (i.e. non-Sienese) recruited from an Italian city with which Siena had no feuds. The Podestà's foreign origins, combined with a non-renewable term of six months, were designed to assure that no single faction would seize control of city government. Unlike Florence, podesterial government in Siena proved itself durable enough to last for several centuries, regardless of the regime in power.[9]

The number of ruling representatives governing Siena varied during the centuries before the city lost its freedom to Florence. Some of the more long-lived regimes during Siena's greatest centuries included the representative governments of the Twenty-four Priors, a coalition of wealthy burghers and nobles who ruled from around 1236 to 1271, and of the Nine Governors and Defenders of the Commune and the People of Siena, who ruled from 1287 to 1355. These groups formed the executive bodies of the Sienese government and chose the three hundred members of the larger General Council. Under the rule of the Twenty-Four Priors, Siena enjoyed its last victory in battle over Florence at Montaperti in 1260, which thus forms a watershed period in Sienese history, marking the beginning of Siena's loss of regional power to Florence. The longest-lasting government in Siena was that of the Nine, who ruled Siena during its period of greatest economic and cultural fluorescence. The Nine was a body of rotating members selected by lot from burghers and nobles of the city. Like many governments at the time, the Nine took a keen interest in every detail of the city's governance and appearance. Through an extensive administrative apparatus they were able to monitor and control the details of policy formation and execution. Many of the larger and most visible structures of the city extant today date from the reign of the Nine. The Nine built major public works such as the Piazza del Campo, the Palazzo Pubblico, and the Mangia Tower next to it. They expanded the cathedral; they legislated the width of the streets; they set design standards for private buildings; they determined the perimeter of most of the surviving walls and gates; and they improved the network of roads and bridges that tied the city to its rural territory. Most importantly, they paid for the discovery and exploitation of the sources of water that feed most of the monumental fountains, many of which were built or rebuilt during their reign. The Nine set in motion the largest single water-supply project, the construction of the *Bottino Maestro* (Main Aqueduct), which extended north of the city almost to the border with Florence (Figure 2.1). As important as performing the initial construction, they maintained the public works they built and vigilantly guarded them against encroachment, appropriation, and abuse by abutters, guilds, and private citizens.[10]

Bringing such ambitious and costly plans to fruition required the city to invest considerable resources in its infrastructure. The Nine were successful in shaping the city's buildings because they could marshal the political consensus and financial wherewithal, establish the administrative apparatus, and attract the technical expertise needed to carry out several large projects concurrently. From an examination of some budget figures from the reign of the Nine, it would appear that the financial resources were a significant, but not determining factor. Using numbers tabulated by Bowsky and comparing them with average annual expenditures on various fountains from the summaries prepared by Fabio Bargagli-Petrucci, we can arrive at figures which suggest that the actual amount of money spent was only a small fraction of the total annual budget of the city. For instance, the entire expenses incurred from the eleventh century until 1299 for building and maintaining all fountains and aqueducts totaled £31,215. By comparison, the average annual budget for the first fifteen years of the Nine, 1286–1300, was about £112,000, with actual budgets ranging from a low of £105,880 in 1292 to a high of £139,439 in 1286. Even though the data are incomplete, one can see that the expenditures for fountains are on the order of less than one percent of the annual budget. Total budgets between 1300 and 1355 steadily increased to annual highs peaking at over £500,000, but the spending on water supply totaled only £87,550 for the entire fourteenth century, suggesting that the water supply budget remained at roughly a constant, minute percentage of the whole. The low percentage of money spent on water supply suggests that costs alone do not explain Siena's role in technological innovation, and suggests that we must look elsewhere, such as the symbolic value of public works or the system by which citizens were rewarded for contributions to the common good, for factors that could have made Siena unique.[11]

The importance and sophistication of the water supply system was out of proportion to the fraction of the budget it represented. To build the water supply system, Siena needed technical expertise. The Sienese obtained the expertise they needed by a combination of paying masters and experts (both native and foreign) and enlisting hundreds of citizens to serve without pay on the dozens of committees that oversaw the details of design, construction, and maintenance. Numerous citizens, not just the ruling elites, served on the various committees that enacted the plans of the Nine. The taxes that financed these projects drew upon every imaginable source of revenue, raising money the way other city republics did, as Bowsky noted, "by taxing everything but water and air." Financial strength, administrative infrastructure, and the availability of a pool of experts allowed the city to defend itself by building strong walls, hiring armies, or paying off advancing mercenaries long enough to focus on domestic concerns.[12]

Remaking the city's infrastructure and improving its appearance required not just money, but also appropriate personnel. The presence of numerous potential patrons other than the city, such as the cathedral and hospital, helped to attract and retain personnel with the necessary skills to shape the city and to draw desirable citizens to it. In this way, Siena resembled the Florence of a slightly later period, which Richard Goldthwaite studied in *The Building of Renaissance Florence.* The fortuitous presence of other institutions, including the cathedral, the hospital, the university, the Dominican and Franciscan preaching friars, at least a dozen monastic foundations, as well as numerous private fortunes allowed master craftsmen in Siena to find frequent—if not completely steady—employment. Such opportunities made Siena a magnet for the ambitious and talented people in the region. Siena's location on the via Francigena, the major road to Rome, facilitated communication and transport with neighboring cities and its own domains.[13]

The presence of multiple patrons made the supply of specialized work that master builders performed more stable. Besides the city itself, another important patron was the Cathedral Works, an institution older than the government of the Nine, dating from the middle of the thirteenth century. Cooperation marked the relationship between the Nine and the Cathedral Works. Both saw their missions as complementary, and thus they shared resources and expertise. The city funded the cathedral, while the cathedral often functioned as a de facto public works department and consulting engineer for the city.[14]

Another major patron of artists, architects, and engineers in Siena was the *Ospedale* (Hospital of Santa Maria della Scala), second only to the cathedral in its antiquity, having been founded in 990. Run by lay brothers and overseen by the city government, the Ospedale became a rich and powerful institution. As its endowment grew, its mission broadened to include providing hospitality to pilgrims, sheltering orphans, feeding the poor, and providing a hospital and hospice during the plague. Performing these many functions required the hospital to build its own vast infrastructure including the hospital and hostels proper, chapels, and other structures within the city as well as several vast farms in the countryside with their own facilities for the storage, milling, and transport of grain. The design, construction, and ongoing maintenance of dams, millponds, canals, and huge milling and storage facilities provided employment for generations of people who possessed the specialized skills the Ospedale required.[15]

The *Studium,* which eventually became the University, was likely to have had more impact on the sort of personnel that the city attracted—especially doctors of law and medicine—than upon the physical structures it built. The foundation of the university predates the government of the Nine by several decades, dating to the 1240s. The city early recognized the prestige and economic advantages a university could bring to it. In its attempt to attract scholars, the city offered them special privileges such as exemption from the evening curfew. Bowsky notes that the Studium of Siena differed from the "university of students at Bologna or of master at Paris" by its secular foundation and governance, allowing it "to break new ground" without treading "through a thicket of Church rights and prerogatives." He adds that the "commune paid professors' stipends, arranged their contracts, and rented the houses in which instruction took place." One of the people who benefitted by working for the university in Siena was native son Mariano di Jacopo, called Taccola (1385–1453). He wrote and illustrated his technical treatise *De ingeneis* (1433) and parts of *De machinis* (1449) while employed as the *Camerarius* (secretary) to the *Domus Sapientiae* (House of Wisdom), a charitable institution that provided housing for poor scholars as well as visiting dignitaries. Not only did he enjoy a sinecure that gave him time to write, but his position also allowed him to meet visiting scholars and others who lodged in the house he managed.[16]

In addition to institutional patrons, numerous holders of private fortunes were eager to leave their mark on the city. Some of Siena's richest citizens were rural nobles who had moved to the city; others made their fortunes in the new economy of commerce and finance. Each wanted a suitable urban palace with which to display his position, attract investors to his enterprises, intimidate his enemies, and house and defend his business, extended family, and servants. One feature that distinguished urban buildings of the later Middle Ages was their greater durability and fire resistance. Stone and brick replaced wood, tile roofs replaced thatch. Goldthwaite states that the buildings of fourteenth-century Florence were more expensive and more permanent than those which they replaced. The same was true of Siena.[17]

Building structures so that they would last required an assortment of master builders and experts as well as skilled and unskilled members of all the related trades that extracted, hauled, and refined building materials and performed specialized construction labor. Goldthwaite makes a persuasive case for the impact the building boom in Florence had upon the city's economy and culture, arguing that construction itself did as much as any factor to shape the Florentine economy and to set the stage for Florentine architectural innovations. It is easy to imagine the construction boom under the Nine having a similar impact on Siena, once one accounts for differences in systems of patronage in the periods under examination. The numerous quarries, docks, workshops, and construction sites not only provided employment, the apprenticeship system by which labor was structured provided training for the next generation of master builders.[18]

To live in a city filled with construction sites, public and private, could not help but attract the attention of citizens outside the industry, if for no other reason than the inconvenience construction dirt, work-site noise, and obstructions to traffic would cause. The impact upon those who took an interest was even greater. A glance at the notebooks of Taccola and fellow Sienese artist, architect, and engineer Francesco di Giorgio Martini (1439–1501) reveals dozens of pages addressing the problems of finishing construction materials, handling them, transporting them, and placing them high above street level. Only against the backdrop of a city under construction do such drawings begin to make sense as practical solutions to pressing everyday problems in the building industry and not just idle speculations of inventive minds.

Drawing on the experience of engineers and artists, the Nine proclaimed the city's increasing political, economic, and demographic strength under its rule by changing the face of the city, its invisible infrastructure, and the laws and financial mechanisms by which it was governed. Like Roman emperors in antiquity, the Nine saw the physical state of the city as a reflection of their ability to govern, and thus they took every measure to see that they looked good. One of the Nine's first acts was to provide for a public palace in which they would both reside and direct the government. More significant for understanding the goals of the regime was its injunction that the other buildings—all private palaces facing the public palace—reflect and repeat its very particular design elements, especially the tri-foliate Gothic window openings. The Nine's continued patronage, combined with that of the private and religious sectors, ensured that many highly skilled masters would make their home in Siena. Individually, they would have a significant impact on the built environment; in consort, Sienese citizens like Taccola, Francesco di Giorgio, and Vannoccio Biringuccio (1480–1537) would eventually create what Paolo Galluzzi has dubbed a "culture of machines," one which would eventually transform Europe by laying the technical, social, and cultural foundations for the Industrial Revolution.[19]

Siena's meteoric growth was followed by a precipitous demographic collapse. The plague of 1348, in what Yves Renouard called an "hécatombe urbaine," wiped out at least half and, according to one fourteenth-century chronicle, as much as three quarters of the city's population between June and August of 1348. Bowsky judges Agnolo Tura del Grasso's estimate of over eighty percent to be high, but he does not dismiss it. Brigands and mercenary armies, such as those of Fra Moriale in 1354, marauded throughout the countryside, interrupting agriculture, commerce, and communications, as well as draining the city's coffers. William Caferro argues that the rising cost of maintaining the peace became a key factor in the city's decline, saying, "of the many calamities that befell Europe in the fourteenth century, there was none worse than the raids of the great mercenary companies." Caferro further argues that at the same time mercenaries were costing Siena so dearly, Florence had found ways to profit from their increasing role. First, the city encouraged mercenaries to spend money acquired in Sienese territory in Florence. Second, Florence created the Bank of St. George specifically for mercenaries to deposit and borrow money acquired at the expense of the Sienese. Thus,

Florence increased the rate at which Sienese finances declined relative to those of Florence, by enabling the transfer of funds from Siena to Florence. Caferro's emphasis on the military is born out by contemporary documents: the chronicles give overwhelming attention to military matters, and the expenses of paying off or employing mercenaries dominate the financial records. Analysis of these records shows that these costs increased as a result of the plague, as Siena struggled to maintain order. Political revolution followed close on the heels of demographic collapse as economic upheavals redistributed wealth after the plague. Bowsky suggests that the impatience of newly empowered citizens with the old regime of the Nine—which had so long kept them in check—was a key element in the revolution that ousted them. He argues that "the Black Death did not directly precipitate the overthrow of the IX. But it was instrumental in creating the demographic, social, and economic conditions that greatly increased opposition to the ruling oligarchy." The Nine fell on 27 March 1355, in a popular uprising that broke out during a visit of Charles IV, seven years after the Black Death.[20]

The fall of the Nine ended Siena's longest period of political stability and began a period marked by increasing factionalism, revolution, and submission to tyrants, both local and foreign. The Twelve Governors and Defenders of the Republic succeeded the Nine and "ruled uneasily for thirteen years." Four revolutions marked the year 1368, out of which emerged the "Reformers." Such frequent changes in government resulted in the institutionalization of factions as political parties called *Monti.* The fall of the Reformers resulted in bitter retribution as 4000 artisans were exiled. The weakness within the republic resulted in weakness to outside threats as well. To avoid a Florentine takeover, the Sienese submitted to the rule of the Duke of Milan, Gian Galeazzo Visconti (1351–1402), in 1399. The power vacuum and the political instability that followed made possible the first Sienese signory, that of Pandolfo Petrucci, in which a single overlord replaced all vestiges of representative government. The Petrucci signory lasted from 1488 until Pandolfo's son Fabio was exiled in 1525. Disintegration continued with ten reforms in the following twenty-seven years, culminating in the siege and Florentine victory of 1555. The expense and disorder naturally had predictable consequences for projects such as water supply that required continuity of funding and leadership to span the generations.[21]

Competing economies, especially that of Florence, gradually but inexorably eclipsed Siena. Siena's demographic, economic, and military decline reduced its ability to defend its borders against Florence's ambition to control all of Tuscany. Siena was constrained to depend increasingly on the farmland and pastures in the malarial, depopulated part of its rural domain that lay toward the Tyrrhenian Sea, where the constant costs of repopulating the territory undermined the return on its investments.

Nonetheless, the system of rewards the Nine created, which privileged public service and public works, lasted through the various regimes that succeeded it, at least until the fall of Siena in 1555. Despite numerous changes of government, including submission to several outside lords, the laws and administrative structures that the Nine did so much to shape remained largely unchanged between 1337 and 1555. One of the most striking features of the revolution of 1355 is how much of the governmental apparatus resumed as though nothing had happened. According to Bowsky "Communal records show a continuity of legislative and administrative personnel . . . [and] administrative techniques were not disrupted. . . ." Continuity marked "the nature of the issues they treated, methods of action, and even in the format and composition of the volumes of deliberations," and "Biccherna and City Council records demonstrate continuity in recording and accounting procedures." No better example exists of the momentum of programs begun under the Nine than Jacobo della Quercia's Fonte Gaia. The Renaissance-style fountain he created to adorn the outlet of the last major aqueduct brought the project to completion in 1419—a project the Nine began in 1334. Unlike all other major Sienese fountains, Gaia was not designed for industrial use; the fountain's primary function was decorative. The eighty-year duration of that project shows how long the grand plans of the Nine persisted after their fall. Such longevity testifies to the enduring qualities of the administrative structures the Nine set in motion.[22]

The momentum the Nine's initiatives created can also be seen in the management of the cathedral's construction and maintenance. By the mid-fourteenth century, the city had acquiesced to the physical limits of the cathedral site and to the financial limits of its coffers. By June of 1357, a panel of experts had determined the building was unsound and the Twelve realized it could not finance the repairs necessary, so it abandoned the Nine's plan to build the largest cathedral in the West. However, the city of Siena did not stop improving the cathedral. Instead, within a generation of halting its expansion, the city began a program of decorating the cathedral floor with marble inlaywork, a project that was to last almost two centuries, until 1550.[23]

Siena's demographic collapse, followed by economic disruption and political decline so soon after its period of greatest expansion, resulted in the wholesale preservation of many of the physical structures of that period. Until the population of the city regained its pre-plague levels in the twentieth century, no new buildings or boroughs were needed. The street plan, the major public buildings, the walls, and the water supply system of that period survive largely intact today because they remained oversized for so many centuries.[24]

PUBLIC WORKS

The construction of the various wells, cisterns, fountains, and aqueducts that constitute Siena's water supply system began long before the Nine, in fact before any surviving written records. Most of the largest and most important fountains, Branda, Vetrice, Follonica, Ovile, Peschaia, and Nuova, date from before 1250. Each of them is built into a hillside and connected to the water-bearing aquifer by a relatively short tunnel. They are all located down-hill from the ridge along which the city is built. The location and form of each suggest that they evolved from a pre-existing spring bank, a place where water naturally emerges from a hillside, and that the tunnels follow the source of the water deeper into the aquifer. The supply tunnels, barely big enough for a person to walk through, function as horizontal well shafts, with the major difference being their curvilinear form; they are not nearly as straight as the Roman tunnels. The city's earlier wells and cisterns are more poorly documented than the fountains, because of their antiquity and the increased likelihood that they were private or religious initiatives that would have generated no records. The construction of the Bottino Maestro, begun in 1334, reflects a major departure from prior practice. Rather than a horizontal well shaft, it is a full-fledged underground aqueduct built along a planned route between the sources of water north of the city to a location at the top of the Piazza del Campo (the city's central plaza). Whereas the construction of the shorter tunnels required the excavators to follow the water source deeper into the hillside, the Bottino Maestro required more sophisticated surveying skills as well as the removal of much more earth. In short, it was an engineering project of unprecedented magnitude that would occupy some of Siena's best talents for over eighty years. The result was an unprecedented amenity: the availability of water in the city's center, and not just its periphery.[25]

An examination of Sienese water supply in the Middle Ages reveals an unexpected degree of complexity of design. Applying end dates to each phase of development is problematic because, as a new system was built, it never completely displaced the previous system. For the first few hundred years of the city's existence, the inhabitants depended upon natural springs, wells, and cisterns. This period represents the first phase of Siena's water supply system. The second phase, from at least 1081 until 1334, entailed constructing major fountain edifices like the Fonte Branda, the Fonte Ovile, the Fonte Nuova, the Fonte Peschaia, and the Fonte Follonica. When the city exhausted the possibilities in mining spring banks, it sought a new source outside its walls, thus initiating the third phase of expansion, from 1334 until 1914. The resulting system, the main aqueduct of the Fonte Gaia, functioned like a Roman aqueduct by bringing water from far north of the city.

Cut into the rock and lined with brick only as necessary, its form, however, was more like a qanat, the gravity-fed aqueducts tunneled through the ground that began to appear in the Middle East during the first millennium B.C.E. The location of the new aqueduct gave birth to an unexpected consequence: whenever possible, the abutters to the route of the new aqueduct tapped into it and connected it to feed their private wells and cisterns. How or when this happened remains obscure. All that is certain is that by the eighteenth century these connections constituted an accepted norm. The third phase of expansion succeeded so well that despite the recurring scarcity of water, the medieval system was not replaced until 1914, and then only partially. This fourth phase of Siena's water supply, from 1914 until the present, is represented by a pressurized aqueduct, running fifty-six kilometers and carrying water from Montamiata to the city. One could argue that Siena ultimately suffered from its once avant-garde system; the medieval aqueducts' continued usefulness retarded the efforts to install a modern, pressurized system. In this respect Siena's experience is typical of cities making pioneering investments in the latest infrastructure, only to have a minimally functioning "legacy system" later become the greatest obstacle to upgrading the system. This study will focus on phases two and three, which emerged during the medieval period.[26]

The city's attention to detail in constructing public works generated a long paper trail, much of which remains in Sienese archives. The records consist of statutes, account books, summaries of General Council deliberations, notes from various other committees and commissions, and several unofficial chronicles. The most accessible source for learning the means by which the Nine sought to enact their plan can be seen in the statutes they wrote and the account books they kept. Today, the surviving statutes consist of roughly fifty volumes covering the three centuries from 1250 to 1545. Two of these volumes, for the years 1262 and 1309, have been published. The *Biccherne* (account books recording income and payments) cover many of the years between 1226 and 1555. About a dozen Biccherne have been published over the course of the twentieth century, but most remain accessible only in the originals. Bowsky mined the Biccherne for the years 1287–1355, basing his study of Siena's finances upon them. The 258 surviving volumes of the *Deliberazioni* (Deliberations of the General Council) are actually redactions of minutes of meetings held between the years 1248 and 1808. Gaetano Milanesi published extracts of these in his *Documenti per la storia dell'arte Senese* in the middle of the nineteenth century. Once again, the only living scholar to have mastered them is William Bowsky, who built the arguments regarding the nature of Sienese culture and society in his book, *A Medieval*

Italian Commune: Siena under the Nine. 1287–1355, and in several of his articles almost entirely upon their contents.[27]

The most exciting aspect of studying Siena is that so many of these documents survive for a city whose physical remains are so intact. One exception is that rebels sacked the public palace and intentionally destroyed many records in the revolution of 1355, an event reflected by the absence of numerous documents. Otherwise, the confluence of physical, administrative, and literary remains presents a unique opportunity to study the evolution of urban infrastructure in the Middle Ages. This book can only begin to look at this mountain of material. I hope that by building upon published documents, Bargagli-Petrucci's research, as well as work that has been done in the last thirty years, this book can achieve a new synthesis of some of the best scholarship in political, economic, and cultural history. By teasing out the relationships between urban patronage in Siena and the creation of an environment where master builders could flourish, I believe that this study can make a contribution to the understanding of the "culture of machines."

This brief overview of Sienese history has so far not brought us any closer to answering the question of why Siena was the birthplace and training ground of well-known engineers such as Taccola and Francesco di Giorgio and the point of origin for their notebooks—two of the most important and most copied technological treatises of the late Middle Ages. According to Bowsky, much of the groundwork for administrative systems and civic culture that was to follow the rule of the Nine took the form of establishing a rewards system for artistic and cultural production that was unique among Italian city-states. Bowsky says the artistic climate the Nine engendered favored the production of public art by reserving the highest rewards for those who contributed to the public good. He argues that much of the difference between Siena's and Florence's artistic production can be traced not to the differing size or wealth of the two cities, but to the varying roles of public and private patrons in the two cities. In this way Bowsky explains both Siena's lack of literary production—which he conceives of as a largely private endeavor—and the preponderance of public art and architecture. In Florence, much of the greatest painting and sculpture was commissioned for private chapels; in Siena the best art works were public commissions. It is easy to see how such a reward system would influence the work of masters in the building trades, as well as the emerging occupation of "engineer." The city granted Taccola various sinecures, such as his post at the Studium which gave him time to write his treatises, and the nickname "the Sienese Archimedes." Besides granting Francesco di Giorgio numerous artistic commissions and contracts to improve the water supply, the city appointed him to several official positions, including the position of one of the four *Provveditores* (the supreme magistracy of the city) in 1493 and entrusted some of its more important diplomacy to him. The city bestowed lucrative public contracts and official appointments on Vannoccio Biringuccio, including director of the mint. Considered in light of Bowsky's thesis, the range of problems Taccola, Francesco di Giorgio, and Biringuccio addressed makes perfect sense: they are precisely the technical issues with which the rulers of Siena had been grappling for two centuries and for which they were willing to pay the highest tribute for solving. Through a variety of monetary and social rewards the city encouraged its greatest technical minds to focus their attention upon issues such as the cathedral, bridges, roads, mining, and water supply.[28]

WATER SUPPLY

Whereas today the gravity-fed water supply system of Siena is emerging as typical of towns built on the sedimentary tufa of southern Tuscany, it did not always appear so. At the beginning of this project the medieval water supply system of Siena seemed unique; it appeared that its only precedent was the aqueduct system of Rome and that much of its importance lay in its symbolic links with Siena's medieval Golden Age (before it fell to Florence in 1555) and with Siena's mythical founders, the Etruscans and Romans. The system's greatest historian, Bargagli-Petrucci, first delineated this view in 1906 when he published his monumental history of the fountains and aqueducts. Since then, most historians have disagreed only with the details of his work; few have questioned his assumptions or the way in which he framed his work. This study will attempt to reframe that history and to demonstrate that the system is more typical than Bargagli-Petrucci thought.[29]

Despite the work of Bargagli-Petrucci and his successors, a gap remains in our understanding of how Siena's water supply functioned and, more importantly, its larger significance for historians of technology. This study seeks to fulfill four objectives. The first is to synthesize a wide range of data, gleaned from archaeological, documentary, and pictorial sources, on the subject of water supply in Siena. The second is to present these data to an English-speaking audience for the first time. The third is to place the structures, people, and events of Siena within a larger historical and geographic context. The last is to assess the impact of the water supply system in Siena for historians of technology, urbanism, and the environment. The overwhelming wealth of sources in Siena and the number of other subjects to which water supply is related combine to present researchers with a staggering array of potential topics. I conclude this chapter with an overview of the book.

SCOPE OF THE PROJECT

The literature to date has concerned itself with either the history of Siena, the biographies of artists and engineers, or the internal history of the water supply system. Out of these approaches emerge three major themes: the urbanization that began around the tenth century, the evolution of the water supply system in response to demographic pressure, and the nature of Sienese engineering. This book will synthesize all three themes in order to place the water supply system into a larger context and answer White's question: Why was Siena such a fertile environment for innovation in engineering? It will start in 1226 when the recorded history of the city's major fountains began and end in 1540, the year of the publication of Vannoccio Biringuccio's *Pirotechnia,* which marked the culmination of Siena's golden age of engineering.

In Chapter Two, "Physical and Historical Context," I argue that familiarity with Siena's geology, geography, and ecology is essential to understanding the way in which the city's water supply developed. Though Siena's site at first appears to be unique, it turns out to be fairly typical of many Tuscan and Umbrian hill towns such as Colle Val d'Elsa, Orvieto, and Perugia. Scholars are only now beginning to learn how many hill towns depended upon a complex water supply system in the Middle Ages. The physical system in Siena differs from those in other towns mainly in being better documented, studied, and preserved.[30]

A huge volume of relevant documents from medieval and early modern Siena is preserved in its public archives and published in the various collections mentioned above. I use these documents in Chapter Three, "Description of the Fountains and Aqueducts," to describe the Sienese system of aqueducts and fountains. I supplement the documents with maps, images, archaeological evidence, and field work. I turn from the Sienese evidence to research on the Middle East in order to develop a typology of Sienese aqueducts. I find that the older and newer types of Sienese aqueducts correspond to Middle Eastern spring flow tunnels and qanats respectively.

The description of the supply system provides a foundation for Chapter Four, "Use of Water and its Regulation," which uses examples from the numerous Sienese statutes to reconstruct the use of water in Siena. In doing so, I elucidate the "demand" side of the water supply equation. The numerous demands upon the system placed it among the pre-eminent public works and city services, on a par with the defense of its inhabitants. Its centrality lent water and the architecture related to it a symbolic value. Structures that conveyed, displayed or dispensed water, like other public works, became a focus of civic pride. The evidence is incomplete, dominated by prescriptive literature, but in identifying major themes and issues about the use of water in medieval cities I provide a framework for discussing the role of water in medieval life. Once its centrality is clear, then it becomes easy to understand how some of the city's best minds came to work on it.

Chapter Five, "Diffusion, Antecedents, and Parallels," revisits the history of the fountains from another point of view. Instead of examining the Sienese historical context, I make a broader sweep of the Mediterranean basin and adjacent regions in a search for technical precedents and parallels. I demonstrate how closely tied together the world of the Mediterranean was, making exotic sources of technological innovation, especially in the qanats of Persia and al-Andalus, more plausible. Larger patterns of common engineering solutions emerge from this comparison. Whether there was any sort of diffusion of technological solutions, at the very least it appears that engineers, confronted with similar geology and analogous cultural imperatives, arrived at nearly identical solutions to the need for an urban water supply. They arrived at these solutions, albeit through diverse institutional mechanisms, regardless of whether their training and knowledge from classical engineering traditions was transmitted through the Latin West or the Islamic portions of the Mediterranean basin. The similarity of the solutions at which they arrived suggests that the scale and nature of urban forms were more important than any cultural or political differences.

By bringing together the scholarship on technology in Siena with the social and cultural history of the city, in Chapter Six, "Sienese Engineers and Their Legacy," I provide a new context in which to understand the technological efflorescence of medieval Siena. An assessment of Siena's water supply system reveals that its ultimate significance lies far beyond the city it served. Although Siena's water supply system itself was not physically unique, its place in the city's culture and self-image and the administrative structure that built it were. Siena's city government, which rewarded public works of art and engineering in so many ways, differed from other late medieval governments. Siena's "cultura delle macchine" and the reward system its government created were responsible for accelerating the creation of a new type of technical worker. The versatility of late medieval and early modern engineers was a direct result of the nature of patronage by cities. Siena was like other Italian cities in that it faced technical problems that the cultures in the eastern Mediterranean and the Islamicized portion of the Iberian peninsula had previously solved. Thus, Siena exerted a strong demand for solutions in the international "market of ideas" and may have functioned as a gateway between East and West. For example, Taccola's notebooks can be seen as a compendium of engineering knowledge gained from the far corners of the

Mediterranean world. Technical manuscripts and notebooks by Sienese masters such as Taccola, Francesco di Giorgio, and Vannoccio Biringuccio distinguish Siena from other Italian city republics. By drawing on the conclusions of Bowsky regarding the unique aspects of the Sienese political system under the Nine and its cultural consequences, I identify factors overlooked by previous students of these technical writings and drawings. The resulting material fabric of "Renaissance civilization" illustrates the composite nature of its sources, wherein traditional, medieval, and ancient technologies were supplemented by numerous foreign innovations in a cultural climate uniquely suited to rewarding engineers and artists who designed and invented for the public good.

In my Conclusion, I recapitulate the major themes of this study. I find that Lynn White, Jr., was correct in identifying Siena as a source of technological innovation in the fifteenth century and find much of the explanation in the insights of William Bowsky, a scholar who never did more than mention in passing the engineering works of the Sienese masters. By bringing Bowsky's understanding of the Sienese cultural climate to bear upon the engineers White and others have identified, I bring together two parallel currents of scholarship to understand better Siena's place in the history of engineering.

Chapter Two

Physical and Historical Context

The Sienese people built their water supply system in response to a specific set of geological, economic, and historical constraints. To understand why the water supply system took the form it did it is necessary to examine those constraints, beginning with an explanation of why the city occupies several hilltops endowed with so little water. The selection of the site has its roots in the changing nature of travel between Northern Europe and Rome in the early Middle Ages and in the manner in which way-stations along this route, such as Siena, responded to these changes.

Siena lies between Rome and Florence about sixty-five kilometers east of the Tyrrhenian Sea. Passing through Siena while traveling from north to south in Italy is no longer automatic. Scheduled flights do not serve Siena's small airfield. To reach Siena by train one must change to a small branch line at Chiusi, two hours north of Rome. Siena has neither a port nor access to a navigable river. The easiest way to reach Siena today is the same way Charlemagne arrived: by road. In fact, one may still follow portions of the same highway he traveled on the way to Rome. Although Siena is out of the way today, it was not always so. Siena owes its existence to its position along the most important road of the Middle Ages, the *via Francigena* (French Road) that connected Rome to French cities. Sienese scholar Ernesto Sestan summed up his city's debt to the geography of travel and commerce by asserting that Siena is a "daughter of the road."[1]

Following the road to Siena on foot, on horseback, or in a car is not only the most convenient way to arrive, but also the most instructive as well. Over the course of centuries, the routes followed through the peninsula were not always stable; they responded to changing circumstances. In fact, the road from France to Rome did not always pass through Siena. Travelers sought security on the road by day and in evenly spaced lodgings at night. The elimination of a way station could cause the route to shift to a road with more lodgings. The relationship between a route and the stopping places along it remained one of evolving interdependence. The early medieval administrative center that coalesced around the bishop of Siena pulled the road closer to the fledgling town and the road brought travelers and wealth. Siena was not only a daughter of the road but a mother as well.

Heading north from Rome, after crossing the stream of Orcia as it flows toward the Ombrone river, the road ascends past Bagno Vignone and Rocca d'Orcia, crosses a few more of the Orcia's small tributaries, and descends into the valley of the Ombrone River. The Ombrone drains one of the two watersheds that Siena straddles, encompassing the land south and east of Siena. The other streams, like the Elsa, which flow north and west of Siena, form part of the Arno watershed. Perched at the headwaters of these two river systems, Siena is left without much water of its own.

At Buonconvento, a walled town founded at the confluence of the Ombrone and its tributary, the Arbia, the road follows the Arbia to the north. If the weather is clear, one can see the outlines of Siena's tower-studded skyline from the villages of Monteroni d'Arbia and Ponte a Tressa. Two towers in particular can be distinguished from all the others. The brick tower to the right, the Torre del Mangia, with parapets jutting out at the top to accommodate a lookout platform and a bell, has marked the seat of civil government since its completion in 1348. Its companion tower on the other side of the Campo, decorated with alternating stripes of green-black ophiolite and pinkish-white marble, is the campanile of the cathedral. The striped tower has marked the seat of ecclesiastical power for as many centuries. Depending upon the angle from which one views the two towers, one tower may seem taller than the other. However, the difference in height is an optical illusion; they are the same. The equal heights demonstrate the parity of secular and ecclesiastical power when they were built. As the road leaves the Arbia Valley it begins its final ascent to the city. The view of the fields flanking the road is now blocked by a row of buildings on each side. Peering between them, one can see that the buildings are only a single layer deep and that fields and vineyards still abut their backyards, even within several hundred meters of the city walls.[2]

Finally, the road arrives at the Porta Romana, an arched brick gateway through which one enters the walled city. The wooden doors are at least two stories tall and almost as wide. Signs prohibiting the entry of vehicular traffic flank the entrance. Since 1963, when the city realized that the economic recovery from the Second World War brought more cars than its narrow streets could hold, visitors arriving by car have been required to abandon their vehicles outside the walls and to complete their ascent to the center of the city

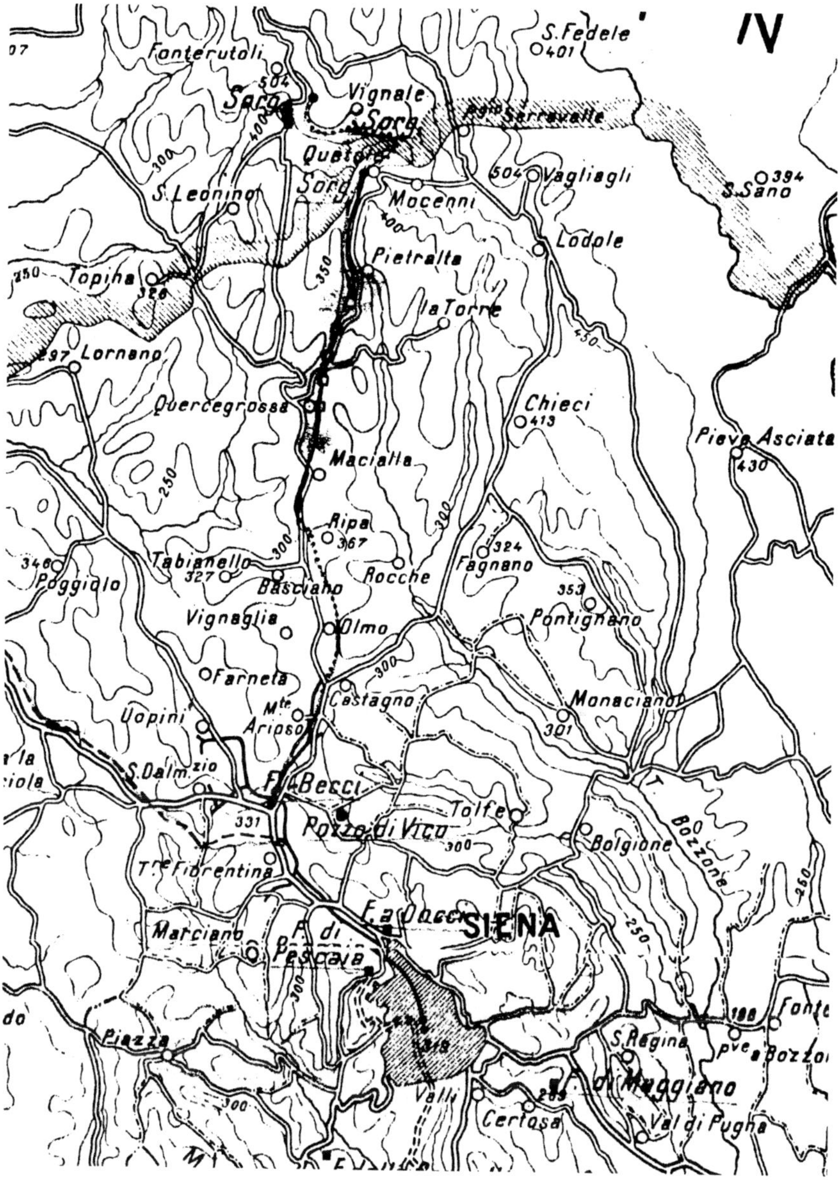

Figure 2.1 Topographic Map of Siena, with main aqueduct extending north from the center of the city toward border with Florence. From Bargagli-Petrucci, *Fonti di Siena.*

Figure 2.2 Skyline of Siena viewed from east. Tower to the left is Palazzo Pubblico, tower to the right is the Cathedral bell tower, the dome is the cathedral. In the foreground is the she wolf. The black and white shield is the *Balzana,* an emblem of the city government. From Battini, *Ricerche.*

on foot. Only residents possessing garages, official vehicles, taxis, and delivery trucks are permitted to drive into the city; even then the hours they may do so are strictly limited. The regulation, based on practical considerations, has the effect of introducing the visitor to the scale of a walking city, a city so compact that the medieval center can be crossed in less than five minutes and whose length can be traversed in less than twenty.[3]

Once inside the city, all vistas collapse into narrowly constrained views of two-hundred meters at most. The streets can be long, but none is perfectly straight. There are no grand boulevards. There are no curbside plantings or trees. There are no sidewalks. The four- and five-story buildings, built right to the curb, keep the sun from reaching the grey basalt paving blocks for most of the day. The narrow road undulates right and left as it wends uphill. After ten minutes of walking from the Roman gate, one notices patches of brightness that draw the eye to the sunlight illuminating a vast open area occupying the physical and cultural heart of the city: the stunningly beautiful Piazza del Campo. The civic tower, which was so prominent from afar, rises eighty-eight meters above the southern edge of the shell-shaped piazza. Directly ahead, behind, and above the buildings along the piazza's northwest edge, are the campanile and dome of the cathedral. The surrounding buildings radiate a pervasive sense of antiquity, and their facades articulate gently repeating rhythms. The antiquity is easily explained: the piazza was designed in the late

Figure 2.3 Winding Streets. Even right after a rainfall, the water does not puddle in the well drained streets, 1994. Author.

thirteenth century, and most of the surrounding buildings had been erected by the fourteenth. Searching the surfaces in an attempt to scan the underlying meter takes a little longer. The most obvious factors contributing to the buildings' harmonious appearance are the uniform height, age, and use of red brick along the continuous curve that most of the façades share. The casual visitor may notice the sets of three arched windows separated by columns that punctuate the front of the civic palace; a careful observer will notice that the form is repeated in many of the facades of the surrounding buildings. A closer inspection of the non-conforming buildings will often reveal a palimpsest of the same arched outlines, now filled in with brick where rectangular windows have replaced the originals.

Climbing the tower permits one to obtain a bird's-eye view of the labyrinthine street plan. The large archway of the Roman Gate lies to the left of the sun's noontime glare. The streets along the ridge extending from the Roman Gate to the Campo can be easily traced from above. Red-clay roof tiles, courtyards, gardens, and numerous cul-de-sacs form complex patterns of red, brown, green, and gray rectangles between the streets. Except along the main roads, the buildings extend no further than the brick walls encircling the city, and often not that far. Due south, to the right of the Roman Gate, the ground slopes downward into a valley of green foliage against a background

Figure 2.4 Piazza del Campo, 1996. The Fonte Gaia is at the right edge of the photo. The campanile and dome of the cathedral can be seen in the background. Author.

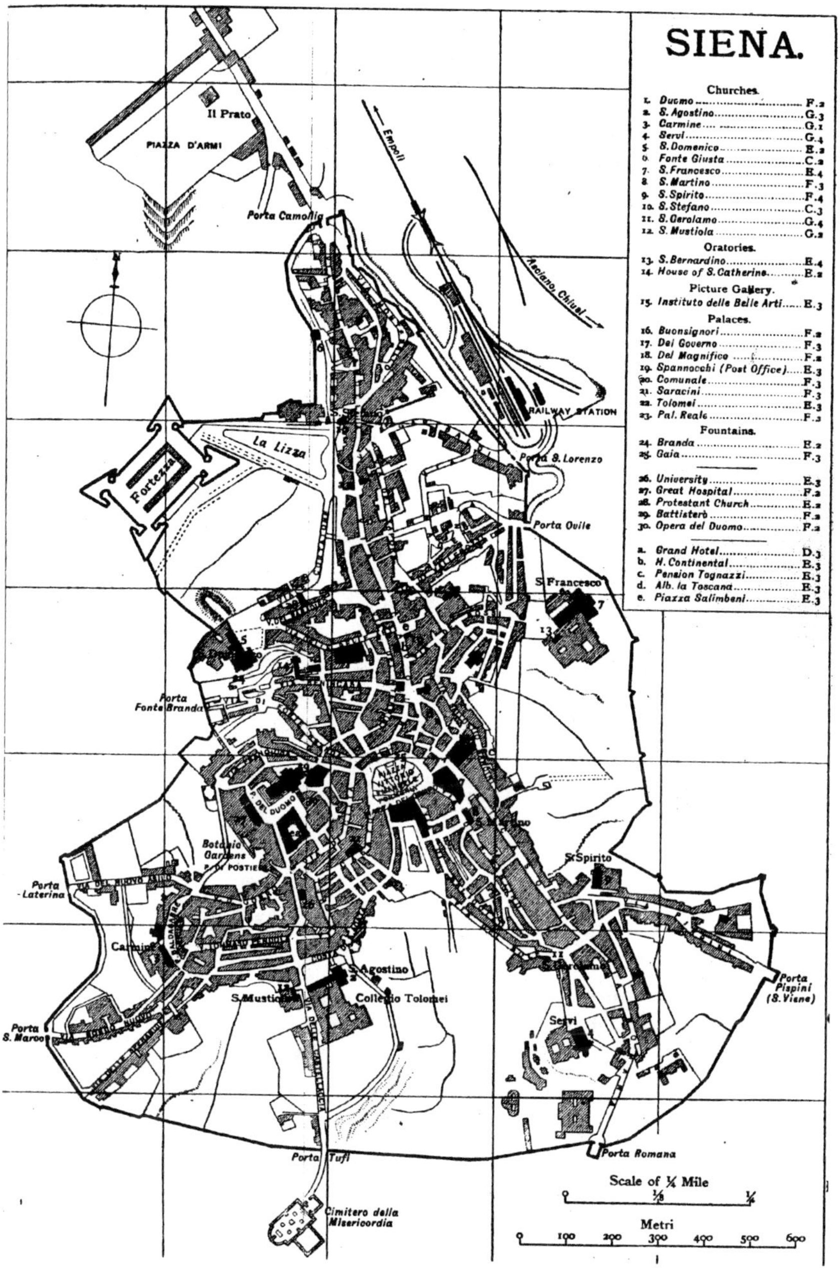

Figure 2.5 Siena's Urban Topography. Note how the buildings cling to the road and how large, unbuilt areas remain inside the walls. Cf. Figure 4.1 for a view of gardens within the city walls. From Douglas, *Siena*.

of brown soil. In the space between the southern wall of the city and the civic palace lies a large cultivated area, which is divided into garden plots and small vineyards. Uphill from the gardens and closer to the back of the palace, is a parking lot for city employees. In the middle of the parking lot is the oval-shaped roof of a nineteenth-century public market. A little further to the right the city extends once again to the limits of its walls. The city is more built up along the roads exiting the walls than in between. Outside the southwestern gate one can see traces of the road that wends through these hills, west by southwest, to the once-malarial swamps of the Maremma, the land adjoining the coast. To the west lies a low, worn mountain range, perhaps fifty kilometers away. This region is called the *Colline Metallifere* (Metalliferous Hills) and has been mined since the time of the Etruscans.[4]

To the north, the land begins to rise, gradually forming the tallest hill within the walls, the part of the city called *Castelvecchio* (Old Castle, the nucleus of the old city), where traces of old walls have left the streets even more tangled and narrow than in other sections of the city. North of Castelvecchio is the hill of Saint Mary, or Colle Santa Maria, the top of which was leveled to form the piazza upon which the cathedral and campanile stand. The front of the cathedral faces west, toward the entrance of the ancient hospital, the Ospedale di Santa Maria della Scala. Founded as a way-station for pilgrims nine hundred years ago, the hospital evolved into a complex charitable institution in the late Middle Ages. Besides housing pilgrims, its most famous function was as a teaching hospital, and it continues in that role today in a sprawling new facility located well outside the walled city. North of the cathedral complex the land continues uphill. The city extends along the ridge that proceeds from the Campo to the north; north is the only direction in which the land rises as it recedes into the distance. The higher land, formed by a spur extending from the Chianti hills, is more forested than the land in other directions. The road from Rome passes north of the city following the higher ground along this ridge. It is this same ridge that ultimately divides the watersheds of the Ombrone and the Arno.

In addition to the gardens south of the market, there are other patches of green within the city walls. The final circuit of walls, dating from the 1320s, seems to enclose too much area, as though it was built in anticipation of growth that never occurred—which is what happened when the plague decimated Siena's population. Florence, by contrast, has very little open space left within the confines of the last pre-plague ring of walls, built between 1284 and 1333. In the foreground, beneath the hill upon which the church and cloister of San Francesco stand, horses graze in a meadow. Beyond the valley of the Chiana rise the Apennines. They form a mountain chain extending from Genoa, along the spine of the Italian peninsula, to the Adriatic coast. A little further to the right, once again, is the now familiar Roman Gate, marking the completion of the panoramic survey.

From these images one can form an abbreviated conceptual framework of the city. The walls roughly delineate an isosceles triangle. At the lower right is the Roman Gate. At the lower left is the gate to the Maremma and the Tyrrhenian coast. To the north is the Camollia Gate, by which the Roman road exits the city. Several kilometers north of the gate the road splits into two. Both branches thread their way through the Chianti hills and across the Arno. Both roads lead to France. One goes through Florence, the other through Lucca. At the northern edge of the Campo, these three important highways converge. The junction of these roads to Rome, France, and the Tyrrhenian coast provided the basis for much of the city's success as a market and administrative center during the Middle Ages.

The three hills, each with its own urban nucleus, combine with the junction of three roads to set a pattern of dividing the city into *terzi* (thirds), instead of the quarters typical of rectangular cities. Multiples of three resonate in the city's statutes and administrative decisions: committees often consisted

Figure 2.6 La Lupa, the bronze she wolf atop the column in front of the Palazzo Pubblico. From Rusconi, *Siena.*

of "six good men, two from each third." Governments have been called variously the "Nine," "Twelve," and "Twenty-four," after the number of elected officials required by their respective constitutions. Multiples of three underlie many of the city's physical structures as well. For instance, eight radii of basalt paving stones divide the pavement of the Campo into nine brick wedges. A statue of a she wolf suckling the twins Romulus and Remus atop a granite column still marks each third of the city. Perhaps, like Athens, Siena began as three villages connected by paths. Like so many artifacts of the Middle Ages, the city's division into three parts continues today in custom and spirit, if not in law.[5]

Besides the cathedral dome there are other religious structures inside the city walls, including two large, boxy churches. The voluminous barn-like brick one, perched along a cliff on the northwest edge of the city is the church of San Domenico. The church of San Francesco abuts the walls to the northeast, above the sloping hill upon which horses often graze. In addition to these capacious sanctuaries of the Dominican and Franciscan friars, dozens of bell towers designating smaller convents, monasteries, and parish churches rise throughout the city. Scattered between them are remnants of truncated stone towers that once served to defend the palaces of Siena's most powerful families.[6]

Figure 2.7 Drain of the Campo, looking toward Palazzo Pubblico, 1996. Author.

The most singular feature of the entire city lies in the immediate foreground, the Piazza del Campo itself. The herringbone-brick wedges of the Campo pavement slope downward and converge on a point in front of the Public Palace. At the base of the palace a large scallop shell carved of travertine directs rainwater from the piazza into an underground drain. From there it flows beneath the Civic Palace and is discharged below the old market into a creek that flows under the city walls.

What cannot be seen so well from above are the dozen or so fountains that supplied medieval Siena with the water that made such a hilltop settlement possible. Today, just as then, the fountains are fed by gravity through underground aqueducts. Therefore, most of these conduits are out of sight, hidden in the valleys which extend below the central ridge of the city. Only the white marble basin of the Fonte Gaia, adorning the upper edge of the Piazza del Campo, is clearly visible from above. Even better concealed than the fountains themselves is the geologic substructure that supports the city and through which flows the city's water supply.

The ridge upon which Siena rests is all that remains of a vast Pliocene sea bottom, composed of various layers of sedimentary materials eroded from the Apennines and deposited about three million years ago. Since then, glacial melt and subsequent erosion have carved the river valleys leading from the Apennines to the sea. The uppermost stratum is a calcium-rich sediment, which over time has been transformed into a slightly more solid rock called tufa. Tufa is a yellow-brown cousin to limestone, porous, fine-grained, and often strong enough to build upon or to quarry as a building stone. Beneath the tufa is an impermeable barrier of clay. A looser aggregate of coarse and fine materials intrudes among the more homogeneous layers often enough to make the results of building construction and the excavation of tunnels less than completely predictable. When rainwater reaches the clay after percolating through the tufa, it is forced to change its course and flow sideways, along the surface of the clay. If the layers of sediment have been carved away to form hills and ridges, then springs will form at those places where the flowing water reaches an exposed edge. This is exactly what happened around the hills of Siena. Such flowing water probably first attracted wildlife, hunting parties, and then settlers to the site.[7]

The water that flows through the tufa comes from rain as snow in Tuscany is rare at Siena's altitude. Most rain in Siena, as in the rest of the Mediterranean basin, comes in the winter. Cool, wet winters followed by hot, desiccating summers characterize the Mediterranean climate. Each year, slightly less than a meter of rain falls in Siena, most of it between October and May. Rain can take several months to trickle along the surface, percolate through the ground, and flow to the rivers that take it back to the sea. The driest months are August and September. In bad years cisterns (which collect rainwater), wells, and even springs can go dry. The summer heat heightens the effect of the annual drought cycle. Native plants long ago adapted to the rhythms of precipitation by adjusting their annual growth cycles or by developing extensive root systems. Early inhabitants responded to the climate by developing transhumance, whereby herders lead their flocks up and down the mountainside with the seasons to follow the forage. But a settlement needs water twelve months a year. The long, wooded ridge, upon the southern tip of which stands Siena, offers a small, but perennial water supply, allowing it to support a permanent population year round.[8]

No doubt this water supply attracted the first settlers to Siena's hilly precincts, but no one is sure when humans arrived. When Romans conquered Etruria during the third century B.C.E., the residents of Etrurian cities became *socii* (allies) who enjoyed a lower status than citizens of Rome. A city named Saena Julia was probably granted Roman citizenship as a concession for ending the hostilities known as the Italic War (or Social War), around 90 B.C.E. Sienese tradition holds the site of Saena Julia to be the hill now called Castelvecchio, the oldest part of the walled city, in the division called the *Terzo di Città* (City Third). Some historians believe that an Etruscan settlement had already existed on the easily defended site. Saena Julia became a Roman military colony under Augustus in 27 B.C.E. Despite these threads of data, much is still unknown. Whether Saena Julia stood on the same site of medieval Siena or was merely located nearby is still being debated. The discussion of Siena's foundation is colored by the desire of many Sienese historians to claim the oldest possible origins for their city. The idea of Etruscan and Roman forebears hold a strong iconic power very appealing to Tuscans, so claims of antique origins of cities and heroic ancestors need to be regarded with caution.[9]

Major gaps in the documents prevent the reconstruction of a continuous chronology. From 309 until 465, for example, no record exists documenting who, let alone where, the bishop of Siena was. In addition, the names of the bishops between the years 465 and 635 are unknown. Lando Bortolotti suggests that Saena Julia, or at least the bishops there, disappeared in the face of the Lombard invasions of 568 and that from the fourth to the seventh centuries there are no documents relating to events in Siena whatsoever and that for the next three centuries they remain extremely scarce. The record of the line of succession becomes more or less continuous only after King Rothari's (636–652) restoration of the bishop of Siena in the mid-seventh century. Although the lack of records alone does not necessarily discredit assertions of

Etruscan or Roman origins, claims based upon the spotty written evidence seem all the more tenuous in the absence of corroboration from archaeological sources.[10]

Numerous relics from Neolithic, Etruscan, and Roman cultures have been found in the area around Siena. A 1979 map catalogs more than thirty archaeological sites within a twenty-kilometer radius of Siena. However, few artifacts have been found within its walls. Although the city has never been the site of formal, controlled archaeological excavations, numerous public works and construction projects over the years have randomly disturbed the soil under the city without revealing so much as a single Roman brick. Remnants of portable artifacts and commodities such as a few potsherds, several squared blocks of marble, and other fragments of undocumented provenance, divorced from archaeological context, offer slim evidence of Etruscan or Roman occupation. They could have been brought to the site at any time in the city's history and thus prove nothing.[11]

Giovanni Cecchini, director of the State Archives of Siena in the 1950s, made an ingenious attempt to resolve the incongruities between the documentary and physical evidence. He suggested that the Saena Julia mentioned in various Roman documents and indicated on a thirteenth-century copy of a Roman itinerary called the Peutinger Tafel was located in one of the river valleys to the southwest of Siena, along one of the known Roman roads. He argued that the inhabitants could have been forced to abandon the site and to move to the present location in the early Middle Ages by the same environmental and political forces that had rendered the towns and roads in the lowlands so dangerous.[12]

The factors making the coast uninhabitable and coastal routes difficult to traverse included not only malaria but also attacks by pirates and foreign armies. For example, Fatimid soldiers based in Africa sacked Roselle and cities on the Ligurian coast, including Genoa, in 935. Pisa sustained successive attacks in 1004, 1011, and 1012. Using newly established bases in Sardinia, Spanish Arabs attacked the coast between Pisa and Genoa in 1015. Cecchini's hypothesis could provide the means for resolving the debate over Siena's origins. It is unlikely, however, that an investigation of sufficient scale to test it will be launched anytime soon. Meanwhile, the current director of the Sienese archives, Carla Zarrilli, is more succinct in her appraisal of Siena's Roman origins; in 1994 she told a Dutch researcher that the idea was ridiculous. The consensus among foreign scholars is that the medieval and modern acceptance of Siena's Roman origins is more important for understanding Sienese perceptions of themselves and their city than any fact supported by the surviving evidence.[13]

The frequency of Siena's appearance in the written record increases again after the eighth century. Siena is noted as a stopping place on the road to Rome in royal itineraries, the most famous of which recorded Charlemagne's journey in 800. A document from the archives of the monastery of San Salvatore called the road from Rome to Paris the *via Francesca* in the year 876. Ecclesiastical leaders occasionally mentioned Siena in accounts of their journeys to the Holy See. For instance, an itinerary of a journey by the Archbishop of Canterbury, Sigerico, mentions a *submansio* (stopping place) at Siena sometime between 990 and 994. Other travelers who mentioned passing through Siena include St. Dunstan of Canterbury (944–988) and Roman Emperor Henry III (1017–1056). In 1059, a conclave held in Siena's old cathedral elected Pope Nicholas II (d.1061). A local tradition holds that one of the routes to the First Crusade passed through the city and that a thousand Sienese were supposed to have joined. Whether or not Siena sent a force to fight in the First Crusade, such traditions nonetheless suggest that the Sienese recognized that the road played an important role in the life of the city.[14]

Evidence of Siena's existence as a physical place, not merely as a literary or legal artifact, begins to emerge in eleventh-century records. By 1070 the walls of Siena had encompassed not only the Castelvecchio section of the city but also the lower hill immediately to its north where the "old" cathedral stood. The inclusion of the cathedral grounds gave a figure-8 shape to the outline of the city's walls by connecting zones around the two hilltops at their tangent points. The next major extension of the walls reached north to encompass the buildings that had been overflowing from the city gates along the via Francigena. The three roads that joined at Siena provided not only a good reason for the city's foundation, but they also continued to shape the direction in which it expanded. Sienese businesses—such as inns—that depended on travelers, sought locations along the major roads. The desire to abut the road resulted in a sort of sprawl, taking the form of an increasingly attenuated, inverted "Y," with buildings extending the city along the roads that followed the three ridges out of town.[15]

Siena had grown large enough by 1145 to cause al-Idrisi, the geographer of Roger II of Sicily, to describe Siena as "a populous center, endowed with markets, artisans, and riches." By comparing Idrisi's description to that of other cities with better-known populations, Lando Bortolotti interprets "populous" to mean that Siena was already home to about fifteen thousand people, a sizeable figure in Latin Europe of the twelfth century. The seemingly high figure is quite plausible when one considers that by the time al-Idrisi wrote his description the walls of the city already enclosed almost half the land that was ever to be built upon. It seems that fifteen thousand inhabitants would have fit quite well within the physical extent of the twelfth-century city. By the middle of the twelfth century the city contained almost thirty churches and religious houses, which would average out to five hundred parishioners and clerics each, not an unreasonable size for a congregation. By 1300, the Sienese walls enclosed about half the area of Florence, leading William Bowsky to infer a population of almost fifty thousand by that year. The population peaked in the years leading up to the Black Death of June 1348 perhaps reaching fifty-two thousand by 1328, not including ten thousand to fifteen thousand "poor, transients, and religious," for a total approaching seventy thousand inside the walls. Construction of new walls to enclose new houses continued through 1346 in all three sections of the city. Despite political upheavals and large fluctuations in population, certain aspects of Siena's urban geography have not changed. The city has remained a regional market, a stopping place for pilgrims, and a center for artisans during the millennium since al-Idrisi wrote his description.[16]

Numerous religious orders founded convents, monasteries, and churches in Siena. As already noted, the Franciscans' and Dominicans' ambitious construction programs resulted in churches that can contain thousands of worshipers. The combined efforts of all the orders have had a major impact on the face of the city. The best known of these orders are the Dominicans, Franciscans, Humiliates, Augustinians, Carmelites, and Servites. In addition to the hundreds of monks, nuns, and clergy who dedicated their lives to the church, several tertiary orders—religious auxiliaries composed of lay brothers and sisters—populated Siena. Such tertiary orders were able to attract some of Siena's most talented citizens. For example, Catherine of Siena (1347–1380), patron saint of Italy, was a Dominican tertiary. The Sienese engineer, Taccola, was active in the tertiary orders of both the Humiliates and San Tommaso. Kees van der Ploeg argues that in Siena, "seemingly even more so than in other cities, an unprecedented religious fervor of a mystical and ascetic character developed in the fourteenth century." Bowsky concurs, writing, "Siena has been called the Tuscan home par excellence of the disciplinant (flagellant) confraternities." The contributions of such volunteers made many of the Hospital of Santa Maria della Scala's programs possible. Religious orders had a profound effect, not only upon Siena's spiritual and intellectual life but also upon its urban form.[17]

Some of the city's most famous and long-lived institutions have their origins in the services provided to travelers. These include the hospital, cathedral, banks, and hotels. The provision of food, lodging, horses, and pack animals depended upon the agriculture of the countryside. Meeting these needs brought great wealth to Siena. Evidence of this vibrant medieval economy's enduring effects can still be seen in the structures this wealth built as well as in the customs of the city. The symbolic importance of raising horses in Sienese territory continues to be demonstrated twice each year in the ceremonies leading up to the Palio, the horse race in the Campo. Although the increased profits realized from serving the needs of travelers, merchants, clerics, and pilgrims were central to the growth of Siena's economy, the acceleration of travel and commerce comprised only one aspect of a period of economic expansion that swept through Europe beginning as early as the eleventh century.[18]

The application of water power to manufacturing processes, especially in the textile industries, was ultimately to prove an important source of wealth and means of diversifying economies that had depended upon commercial and financial services. However, this new economy of water-powered production put hilltowns at a disadvantage. Bowsky says of twelfth-century Siena that its "lack of a major source of water condemned it to lag industrially behind the increasingly populous and industrialized city of the lily." The full potential of water power would not be realized for another two centuries. When Siena's wool production was at its peak in the 1330s and 1340s, Florence's production was at least eight times greater, at one hundred thousand cloths compared with Siena's nine thousand or ten thousand. Towns built along rivers that could be dammed and channeled to power waterwheels would emerge as the greatest beneficiaries of these innovations. Milan, Bologna, and Siena's foremost rival, Florence, were just a few of the towns that built fortunes on an industrial base beginning in the eleventh century. Carlo Cipolla notes that already in 973 there was a fulling mill at Parma, with others following at Verona in 985, Florence in 1062, and Prato in 1107. Water provided the power for paper mills in Xativa, Valencia, by 1151, in Genoa by 1210, and in Fabriano by 1230.[19]

Sienese merchants, bankers, and manufacturers were not unaware of this new source of power. They tried to exploit Siena's limited water resources to their fullest, but Siena had only enough power within its precincts to turn a few small mill wheels, mostly used for grinding flour. The application of water power favored cities along the fall-lines of rivers. Without a river like the Arno that flowed through Florence, the Reno that flowed through Bologna, or the tributaries of the Po that powered industries in Milan, Siena could never apply enough water power to the production of textiles to keep up with its competitors. Siena's prosperity, primarily derived from its function as a regional market and banking center, was to be eclipsed by Florence's industrial wealth, a pattern that was to be repeated in numerous hill towns throughout northern Italy.[20]

The economic sea change wrought by the application of water power to industrial processes had differential consequences, depending upon each city's ability to benefit from new sources of energy. Recognition of such possibilities increased the desire for more industrial power and spurred the best minds in Siena to propose means of overcoming the city's lack of water. Beginning in the thirteenth century, Siena went to great lengths to increase the supply of water for both industrial and domestic users as the city's population grew. Its leaders consulted experts from inside the city and from abroad. The General Council entertained suggestions, such as searching for an underground river they called the Diana, believed to run along an unknown route once known to the ancient Romans. In 1295 the General Council even allocated £70 to the chief of the cathedral works to find the Diana. This contract was one of many the Nine let in 1295 to search for water. Despite the failure of these efforts, Siena continued searching for new sources of water to satisfy industrial users and the needs of the city's growing population. The Nine continued their quest for new sources of water until they met with their first large-scale success when the Bottino Maestro brought water into the Campo in June 1343. Despite this turning point, and the plague that would reduce the population five years later, bringing an adequate amount of water to the Campo would continue to be a problem for the next several centuries. Although the Bottino Maestro increased domestic water supply, it did little or nothing for industrial users who needed much more water than it could supply.[21]

In the 1420s Sienese engineers such as Taccola attempted to devise mechanisms for exploiting extant sources of water, as well as to continue looking for new sources of power. For example, he contemplated the construction of vertical-axis windmills, a new and unproven technology at the time in central Italy. Taccola's notebooks also reveal that he looked as far away as Toledo for models of how hill towns could be supplied with water; two of his sketches of water-lifting devices have a city labeled Toledo—a city famous for its gardens, irrigated by large water-lifting wheels called norias—in the background. There is no evidence, however, that any of Taccola's schemes was ever applied to the problem of water supply in Siena. The economic stagnation and decline into a market primarily serving the local countryside foreshadowed Siena's political subjugation to Florence beginning in 1555.[22]

Siena's slide in relative productivity was further complicated by the geography of its rural territory. The process by which Siena's territory took shape had begun when the city launched a suit against Arezzo over its borders in the eighth century. The verdict, in the court of the major-domo of King Liutprand, against Siena ultimately established its eastern borders not far from the city walls. The skirmishes with Florence in the 1140s also set the northern borders of Siena's territory uncomfortably close to the city itself, leaving much of the most useful watershed (because it was at a higher elevation than the city itself) under Florentine control. After Siena's last triumph against Florence at Montaperti in 1260, the series of territorial battles with Florence, stretching from the middle of the thirteenth to the middle of the fourteenth century, ultimately demonstrated Siena's military weakness. William Caferro has argued that roving mercenary companies had a more devastating effect on Siena's economy than previously acknowledged, playing a pivotal role in the city's decline. The resulting directions in which Siena could expand its economic base, with the least potential for political conflict, were west and south, toward the less promising and sparsely populated forests and malarial wetlands along the Tyrrhenian coast.[23]

The pull of mineral wealth complemented the push of expanding neighbors, bringing Siena into the region that disease had made virtually uninhabitable for at least a thousand years. The hills southwest of the city had long been known to contain valuable minerals, but the surrounding wetlands, infested with mosquitoes, made it dangerous to live there, or even to pass through the region. Nevertheless, Etruscan cities such as Populonia had become wealthy by exploiting the mines in the mountains in the Maremma and by processing iron ore from Elba since the middle of the seventh century B.C.E. Possession of a mining region was to provide Siena with not only wealth but also with experts and technical knowledge. The first notice of Siena's gaining an interest in the silver mines to the west appears in 1137 when the bishop, on behalf of the commune, received the fortified town of Montieri and its silver mines, thirty-five kilometers southwest of the city. In 1147 Siena continued its seaward expansion with the acquisition of Montepescali, twenty kilometers further than Montieri. Sometime before 1180 Siena had lost its rights to Montieri, but by 1180 it regained one quarter of the citadel and mines. The silver thus mined was put to good use when Frederick Barbarossa granted Siena the right to mint its own coinage in 1185 or 1186. Perhaps the wealth from silver mines and the advantages of being powerful citizens of a city that operated its own mint contributed to the ability of the *Gran Tavola* (Great Table) of the Buonsignori to become the bankers to the Papacy in the 1220s. In the end, the economic impact of the silver extracted may have been outweighed by the value of the mining skills themselves. Siena's ability to tap the supply of skilled workers and the specialized technical knowledge the miners guarded so closely was to make a lasting contribution to the repository of technical knowledge in the city. Such metallurgical knowledge was to persist as an important dimension of Siena's technical knowledge and cultural life at least until 1540, when native son Vannoccio Biringuccio, who had directed the iron mines at Boccheggiano, published his *Pirotechnia.* Generations of engineers, such as Taccola, Francesco di Giorgio Martini, and Vannocchio Biringuccio, would draw upon the knowledge of miners, smelters, and metal workers to advance their own work, both in the studio and in their treatises.[24]

Likewise, Siena's early resolution of the roles of secular and religious authorities allowed an extensive and fruitful collaboration between church and state in the construction of public works and in the governance of the city. Perhaps the most important result of collaboration in projects, such as the construction of the cathedral, was the transfer of management skills and technical knowledge from religious institutions such as the monastery at San Galgano or the cathedral workshop to the city's secular agencies. Standing atop the Torre della Mangia, peering down at the streets, piazze, and walls, one can only discern the larger patterns and bolder outlines of the city's form. Despite thousands upon thousands of small changes in the urban fabric, from electrification to the installation of flush toilets, by day the city looks not unlike the city of 1350. The evidence that bears witness to otherwise invisible changes in the city's infrastructure, such as the provision of a modern water supply system, can only be read from street level. What the city was like before systematically overhauling its urban infrastructure in the beginning of the twentieth century can only be constructed in the imagination. Thus, the visitor to Siena must return to ground level to understand the physical structures that the Sienese employed to deal with the constraints set by its poorly-watered, hilltop location.

Chapter Three

Fountains and Aqueducts

In 1625 Giugurta Tommasi published what he called an old proverb that said Siena "keeps its beauty underground." Often quoted but unsubstantiated accounts attribute a similar statement to Charles V, who upon emerging from a tour of Siena's *bottini* (underground aqueducts) in 1536 exclaimed, "Siena is more beautiful underground than above." What is it that so captivated visitors about Siena's underground tunnels, carved out of the living rock? This chapter will explain the tunnels' attractive qualities by describing their appearance, location, extent, and fundamental role in supplying the city's fountains with water. It will then examine the ways in which the hidden tunnels complement the visible parts of the water supply system. For certain questions, such as the tunnels' origins, this study can offer only tentative answers.[1]

As noted earlier, all the water feeding the fountains in Siena starts as rain, less than a meter of it each year, falling mostly between October and May. Some of this water runs along the surface, eventually finding its way into small streams, and some percolates into the ground's permeable layers where it is trapped by the clay below, replenishing the aquifers that are Siena's principal sources of water. Since the city's foundation, Siena's residents have obtained water in four distinct ways. They collected rainwater in cisterns. They dug wells into the aquifer. They excavated spring banks and built fountains. And they built an aqueduct to bring water from springs outside the city to its center. These four water-supply systems served Siena for the first thousand years of its history.

Using multiple sources of water conferred several advantages upon the citizens of Siena, foremost among them being increased reliability through redundancy. Dora Crouch, in her study of ancient Greek water management, argues that redundancy of water supply offers increased security in the face of the failure of any one sub-system. She says that such redundancy is a feature of Greek cities in the Mediterranean. The use of overlapping water supplies is likely to be common to all urban civilizations coping with the Mediterranean basin's pattern of unevenly distributed annual rainfall. The perennial threat of siege warfare, during which an enemy might try to sever a city's supply of food and water, also demanded a backup system. Crouch adds that a concomitant aspect of the Greeks' strategy included the categorization and use of water according to its quality, reserving the scarcest commodity, potable water, for drinking alone. We shall see that the Sienese employed a similar system of allocating water to users according to quality.[2]

The Sienese were not the first in Italy to use such criteria in the management of water resources. Frontinus' description of each of the eleven waters brought to Rome by the aqueducts shows a similar system of allocation of hydrologic resources according to quality. He says that the aqueduct called Marcia, "as an obvious starting-point, would thus be reserved entirely for drinking. The others would be assigned to appropriate uses, each according to its particular qualities." Low-quality water had its own uses such as "watering gardens and the meaner uses of the city itself." Frontinus says that Augustus built the Aqua Alsietina only to fill the Naumachia, an artificial lake used to stage naval battles. In sum, the best waters supplied the public drinking fountains, whereas the lowest-quality water was used for public entertainments, irrigation, and flushing the city's gutters and drains. The Sienese embraced a similar hierarchical arrangement for the use of their multiple water supplies.[3]

The lack of detailed evidence about medieval water use prevents us from establishing firm conclusions at present, but it seems likely the ancient three-tiered system of applying water supplies of springs, wells, and cisterns to appropriate uses was in effect during the Middle Ages as well, though probably not as a direct continuation of ancient practices. Until the modern, pressurized water supply system was installed early in the twentieth century, the rainwater collected in cisterns played a much more important role as the tertiary water supply, supplementing water from aqueducts, springs, and wells. Then, as now, the Sienese reserved the best water for drinking. Even in the 1990s, some residential kitchens in Siena were equipped with an additional spigot so that both potable and non-potable water would be available to the homeowner.

The evolution of methods for exploiting water can also be divided into chronological stages, beginning with springs, then wells and cisterns, and finally using aqueducts to extend the city's effective watershed. Springs are naturally occurring sites where water from an aquifer seeps out of a hillside, usually to circumvent an impermeable layer of ground. Since Siena lacks a nearby river and is upstream from the tributaries to nearby rivers anyway, naturally occurring springs provided the earliest sources of water. Using them required no physical interventions; they only had to be discovered and have their potability assessed. Modifications of Sienese spring sites began well before the oldest Sienese documents. Springs were vital to hunters and travelers who would know where they could find water, even in uninhabited regions. Year-round springs were a crucial factor in selecting a site for settlement in the Mediterranean basin. Perhaps intervention began with placing a stepping stone in the waterlogged ground or clearing away the underbrush at a spring. Such measures would likely have been followed by forming a basin into which one could dip a pitcher. The next step would be to construct a retaining wall to keep the surrounding earth from caving into the water. Attempts to increase the flow of water by digging into the spring bank could have followed. The earliest written records refer to fountains and note finding new veins of water, but make no mention of unexploited springs. Thus it appears that by the thirteenth century, when the surviving written record becomes abundant, all visible springs in the city were already recognized and used. That the springs had all been named and exploited by this date is not surprising in that the production of extensive written records and the more intensive exploitation of natural resources are but two of the concurrent results of the city's demographic success in the twelfth century.[4]

We do not know exactly how these spring sites evolved into fountain complexes, but we can construct a hypothetical sequence of events by piecing together various archaeological and textual clues. The earliest documentary source to record construction details related to Siena's water supply is in the volumes of the Biccherna, the ledgers of Siena's treasury, dating from the second quarter of the thirteenth century. The volumes of the Biccherna note the concurrent construction of both fountains and the aqueducts that supply them, suggesting that the process was well underway by 1226. During those same years the city repeatedly extended its walls to enclose its rapidly growing population. By 1300 the city walls encompassed and protected most of the city's fountains as well as its almost fifty thousand residents. The fountains were all located at the perimeter of the city, at a lower elevation than the more densely settled city center, which is why the process of extending the walls to encompass them was such a prolonged one. The most abundant of all these fountains was—and still is—the Fonte Branda. This fountain is built into the largest, and one of the highest, landmasses. The low ridge rising above the fountain extends north of the city. It appears that the Branda's aquifer has the largest recharge area and thus has the most potential for bearing water.[5]

Figure 3.1 Fonte Branda, 1996. Author.

Wells represent the next stage in exploiting groundwater. Although digging a well requires more effort than using a spring, wells allow a greater degree of freedom in their placement. Wells are vertical shafts, often lined with a permeable wall of stone or brick, dug deep enough to intercept the aquifer. The aquifer can occur at almost any depth, from several meters to hundreds of meters below the surface, and can vary greatly in thickness. In addition to geologic factors, the practical limits of digging wells by hand are determined by a combination of the cost of labor and materials and by the length of the rope used to raise the water bucket. Beyond a certain depth the weight of the rope exceeds the weight of water raised, creating a point of diminishing returns. These limits restrict the use of wells to zones where the aquifer is within one hundred fifty meters or so of the surface, depending upon the type of soil or rock being excavated. Most of the documented wells in Siena are located within the confines of large institutions or at major crossroads in the city. In either case, the work would have been distributed among a larger group of laborers than would have lived within the confines of a single household.[6]

Although wells usually furnished potable water, water from wells was comparatively expensive, given their potentially low yield and the labor required to raise the water the length of the shaft. But wells could be conveniently located in the courtyards of houses rather than away from buildings. A well or cistern in a courtyard was more defensible than a fountain or spring beyond the boundaries of one's neighborhood. Especially in the medieval city, where neighbors could present a more immediate threat than foreign armies, a secure water supply was a highly prized feature for any household. The value placed on the availability of a good water supply in the fourteenth century is reflected in Boccaccio's descriptions of ideal Florentine country villas as being endowed with "wells containing the freshest water" or with a spring in the cellar "surging with great quantities of the coldest water."[7]

Shallow, hand-dug wells were especially vulnerable to contamination. In valley-floor cities such as Florence or Rome, wells were subject to contamination from nearby cesspools, from infiltration of salt water along the coast (as at Pisa), or from agricultural waste from herds above the city (e.g., in villages throughout Tuscany). Nonetheless, because they reduced the amount of time spent hauling water, wells were the preferred solution in Siena until spring water flowing through the Fonte Gaia aqueduct became available in the fourteenth century.

To insure that well water would be accessible even to those who did not have their own well, the city offered financial incentives for their construction. The Biccherna from 1262 records reimbursements of at least £15 made to three religious institutions for expenses they incurred digging wells of *acqua viva,* "live water," signifying that they had reached a good aquifer where the water was flowing, and had not merely built a cistern that stored rainwater. The account books give the locations as Santa Agata (downhill from the oldest section of the city), the monastery of the Umiliati (north of the center of the city, on the east side of the road to Rome), and the Servite convent of Santa Maria (in the southeast corner of the city, also on the road to Rome). The net result was the addition of three wells, evenly distributed, one in each *terzo,* or third, of the city, two of which would be especially convenient for travelers passing through the city.[8]

Sometimes a well would tap a particularly abundant aquifer and produce more water than could be used locally. Or changing patterns of land use would reduce the demand placed on a well, creating a surplus. If the aquifer was high enough in elevation, a tunnel could divert it to another outlet. A statute from around 1309 records one such diversion. The statute calls for channeling water several hundred meters from the well of the Umiliati friars to the Fonte Ovile. More common than any local surplus, however, was a chronic shortage of water. The process of paving over an ever-greater percentage of the city's surface exacerbated such shortages by preventing rain water from recharging the aquifer beneath the city.[9]

Figure 3.2 Cistern in a courtyard along via Pantaneto, 1994. Author.

One source of water could increase in proportion to the number of buildings covering the city: cisterns. Cisterns took advantage of increased density of buildings because they collected water from roofs. When more water was needed than a well could supply, or if the aquifer was too deep to be efficiently exploited, one could build a cistern. A cistern is a hole in the ground, often excavated in a courtyard. Its interior walls are lined with brick and made water-tight with a plaster coating. It is usually filled by water from downspouts draining the roofs of the surrounding buildings. The more roofs connected to it, the greater the potential surface area available for collecting water. Since cisterns are fed from above and not by the water table, they need not be as deep as wells. Their storage capacity can be augmented by increasing their diameter, which is usually more cost-effective than digging deeper. Shallower cisterns are also easier to maintain than wells. A person can clean or repair a wide shallow cylinder more easily than a narrow deep one. Cisterns need to be sufficiently large to hold enough rain water to make it through the dry season. In Apulia, the southeast region of the Italian peninsula, roofs and the conduits that directed water to the cisterns were all part of an integral plan typical of medieval and early-modern construction, demonstrating that the redundancy and hierarchical typology of water were features common to several Mediterranean cultures. Cisterns were not a medieval innovation. Greek and Roman cities depended on water provided by cisterns. The use of cisterns has continued through the ages because in some locations no better system has been installed.[10]

Cisterns of a slightly more elaborate design, which included a filtration system, were the sole source of potable water in the Venetian archipelago from the early Middle Ages until the sixteenth century. The lack of an accessible water table under the Venetian lagoon made wells completely unfeasible. Instead, Venetians depended upon a highly developed type of cistern that incorporated a filtration layer of sand. Venetians had potable water delivered in tar-lined boats from the mainland. The sand filter was said to remove impurities from rooftops or even the taste of tar from the delivery boats.[11]

Cisterns functioned both as a source of lower-grade water and as a backup system to sources of potable water on a neighborhood basis, in both the city proper and the settlements that formed along the roads into the city. Like wells, cisterns were usually located in courtyards. Cisterns proved vital not only to the households that built and used them but also to the city as a whole. Together, a city's cisterns formed a defense against drought. Their importance on the collective level can be inferred from the financial incentives Siena offered to persuade citizens and institutions to build them. The commune was willing to reimburse owners £15 in counted coin and ten barrels of lime for each cistern built, an incentive equal to the previously mentioned one it offered to dig wells. In 1263 the city paid the Cathedral Works (the semi-autonomous organization responsible for the physical aspects of the cathedral complex) for a cistern it constructed in accordance with the city statutes. The city offered a similar incentive to the large households which occupied the palaces at the perimeter of the Campo, a heavily used area of the city without its own fountain until 1343. Cisterns holding collected rainwater could supply uses for which absolute purity or freshness was not a prerequisite, such as watering animals, bathing, laundry, or wet mopping. They could provide emergency drinking water during periods of water shortage caused by drought, siege, or the local government's failure to maintain the public water supply.[12]

Replenishing the cistern depended entirely upon enough rain falling each winter. Unfortunately, water stored for six months in a cistern was vulnerable to contamination and stagnation. Both taste and practice among users of cisterns seem to have reflected these risks. Crouch found that when people in the Hellenic world had a choice of stored water from cisterns or fresh water from springs they preferred the taste limestone imparted to the spring water to that of stored rainwater.[13]

Together, water from springs, wells, and cisterns supplied all of Siena's domestic water supply, with perhaps one exception, until the twentieth century. The city twice studied plans to divert water from nearby rivers to the city. The first time was on 13 January 1267, when several visionary Sienese leaders attending a General Council meeting suggested building an aqueduct to bring water from the Ciciano, a tributary of the Merse, along a route that would have stretched sixty-five kilometers from the southwest of Siena. One of the participants, Bartalomeus Saracini, suggested that a Cistercian monk, Brother Agnolo, be employed to survey the route. Ultimately the General Council decided against the project. Nonetheless, the grandiose scale of such a project remains striking. A twentieth-century engineering assessment of the proposal concluded that such a long aqueduct would have cost too much and would not have brought the water above 288 meters. This elevation would have placed the water well below the city center, too low to have been worth the expense, had it been technically possible.[14]

The second proposal, in 1358, called for bringing water from the Staggia River, near the border between Siena and Florence, into the Fonte Gaia. From a technical point of view this project was the most ambitious to date. It remains the most interesting one from a historical perspective because it would have required the construction of a siphon sixteen hundred meters long. A few tantalizing scraps of physical evidence, published as "Relazione intorno alle acque delle fonti di Siena . . ." by Sienese physician and historian Domenico Battini suggest that this siphon may well have been built or at least attempted. Duccio Balestracci speculates that the city attempted to expand its water supply to improve public health. The city's finances and administrative apparatus bounced back quite rapidly after the plague. Perhaps the newly installed government was attempting to satisfy pent-up demand for more water after the revolution of 1355. In any event, it is puzzling that the city would contemplate obtaining water so close to the Sienese-Florentine border at a time of active conflict with Florence.[15]

Battini wrote a brief history of Siena's water supply system to support his case for improving public health in eighteenth-century Siena by increasing the supply of potable water. His "Notizie istoriche sopra gli acquidotti delle fonti di Siena in schiarimento ed illustrazione della precedente Relazione" appeared as a section of the "Relazione" mentioned previously. In his history Battini argues that in the late fourteenth and the first half of the fifteenth centuries Siena had extended the aqueduct of the Fonte Gaia through enclosed pipes to divert several sources of the Staggia River, at points north of the city near Quietole and Stella, but by 1462 at least the acqua del Castagno had already been "lost" for sixteen years. Battini reconstructs the probable route of the aqueducts by using official maps and documents drafted by Giuseppe Fondi (the city official in charge of the aqueducts) in 1702. He supplemented his archival research with visits to the site. Battini's fieldwork confirmed Fondi's observations that traces of tile pipes as well as the stone channel through which they ran could still be found along the route, despite the destruction wrought by years of cultivation and scavenging for building materials. Basing his decision on the combination of Battini's description of terra cotta pipes encased in a heavy masonry conduit and the dip the

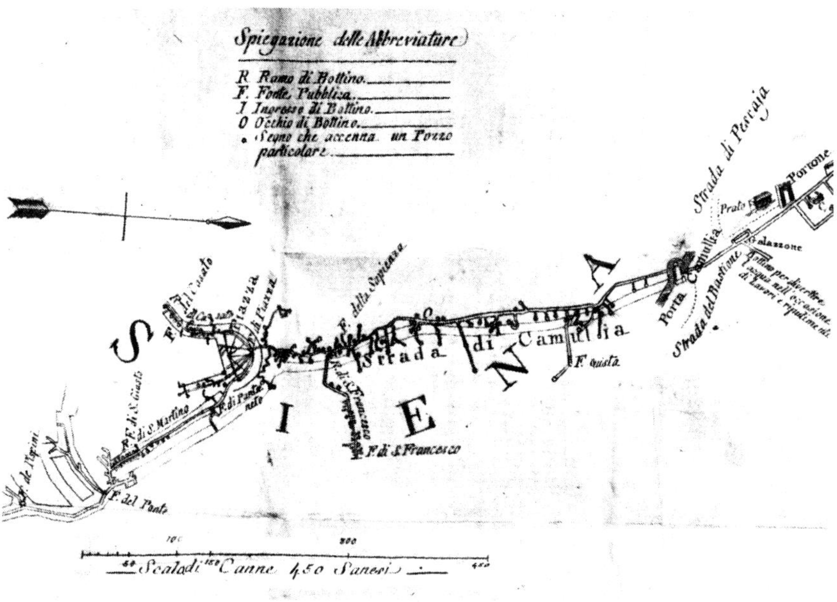

Figure 3.3 Detail of map of the Bottino Maestro and its branches. R. = Ramo (branch), F. = Fonte pubblica, I.= Ingresso di Bottino (entrance), O. = Occhio di Bottino (*smiraglio* or ventilation shaft), o = denotes particular Pozzo (well or cistern) connected to the aqueduct. The Sienese *canna* is about 2 meters. A little to the right of the Porta Camollia, can be seen the Galazzone, the large reservoir and settling pool. From Battini, *Ricerche,* courtesy National Library of Medicine.

aqueduct takes, Bargagli-Petrucci concludes that the clay pipes formed a siphon. If so, siphons' notorious vulnerability to failure could explain why that section of aqueduct was soon abandoned.[16]

Except in plans never brought to fruition, local rivers did not play a role in Siena's domestic water supply because of their distance from the center of the city; for example, the Merse was thirty kilometers distant, the Staggia twelve kilometers distant, and the Arbia nine kilometers distant at their nearest points. The more important, limiting factor was that the headwaters of these three rivers were at elevations lower than that of Siena. The role of rivers in Siena's water supply was, therefore, largely limited to industrial and agricultural uses outside the city walls and in the countryside.

The city of Siena never possessed enough water for more than a few, small industrial users. Tiny, often seasonal streams of water, such as the Tressa, Riluogo, and Bozzone, formed in part by the waste stream of water leaving the city, provided a small and not always reliable supply of water to some mills near the city. The best known of these mills were those below the Fonte Branda, traces of which remained visible on late nineteenth-century military maps of the area. Records of legal battles over riparian rights and notices of years when the mill ponds went dry reveal the inadequacy of the water supply from rivers and streams near the city.[17]

In years of drought the city resorted to extreme measures such as using "dry mills" (whether turned by manual or animal power the documents do not say) or hauling grain by pack animals to mills far from the city, such as those along the Merse, thirty kilometers away. The scarcity of workable mill sites is revealed further when one learns that for access to the meager supplies below the Fonte Branda millers were willing to pay the city an annual rent. For instance, Master Ugolino Bernardini and Ugolino Martini paid one hundred soldi per year in 1262 to rent a mill site next to the *guazzatoio* (horse pool) of the Fonte Branda.[18]

Drought was not the only climatic extreme with which Siena had to contend. Every winter, especially in wet years, rising rivers threatened to wreck mill sites in the Sienese countryside. The dangers of flooding were particularly pronounced along the Arbia, the only river well suited to building mills. A monumental example of these thirteenth-century mills still stands several kilometers south of the city at Monteroni d'Arbia. The scale of the mill and its brick construction have surely contributed to its survival into the twentieth century. The mill's location, in a wide flood plain instead of a narrow valley, allowed flood waters easy passage around its base. The Arbia demonstrated its ability to rise above its banks and sweep away everything in its path as recently as the winter of 1993–1994, when floods destroyed two of the three reinforced concrete and steel bridges connecting Siena with the Chianti Hills. Unlike the mill at Monteroni d'Arbia, they had been built across narrow ravines, a setting that channeled the power of the river against the bridges' piers and abutments.[19]

The lack of trees along the river banks increased the destruction wrought by flooding. Trees and shrubs have the ability to slow flood waters and absorb some of their energy. Although it lacked the modern vocabulary to explain its decisions, the Sienese government recognized the role trees played in stabilizing ecosystems as early as 1309. A law from that year orders that trees be planted south of the city, in the flood plain of Buonconvento every two braccia (a little more than a meter) "to conserve the banks of the road." This law suggests that urbanization, deforestation, intensive cultivation of the land, and overgrazing, each precipitated by late medieval demographic growth, exacerbated winter flood damage early in the fourteenth century. The same lack of trees and vegetative cover would have accelerated summer runoff as well. The resulting decreased absorption of water into the ground lowered the water table, increasing the impact of the annual dry season. The race to increase the food supply for the growing city brought steeper and steeper land into cultivation, which increased the effects of runoff of surface water onto roads, from river banks, and into streams and rivers. This pattern is repeated wherever uncontrolled development, grazing, or deforestation takes place in river basins.[20]

Despite the city's efforts, springs, wells, and cisterns together did not bring enough water to the places where it was needed most. The solution in Siena, before the application of steam power to pumping water in the nineteenth century, was to dig longer, not deeper wells, and to dig sideways, further into the aquifer. This increased the effective surface area of the water-bearing material. The Sienese called these extensions of their springs *bottini.* The very process of following the aquifer accounts for the often winding, curving, organic forms bottini assume.[21]

Before continuing, it is necessary to note that two types of bottini supplied water to Siena. The difference between them is subtle but important. Unfortunately, they have always been lumped together under the same term, bottini, ever since city documents applied this term to the second type of aqueduct, employed to bring water to the Fonte Gaia in the 1340s. I will instead distinguish between the older, Branda type and the newer, Gaia type. In the former the tunnel, with permeable walls and ceiling, is both the source and the conduit. In the latter, the Gaia type, a non-porous channel conveys water from various springs to the fountains. The Gaia type is essentially a qanat, a form that will be explained below.

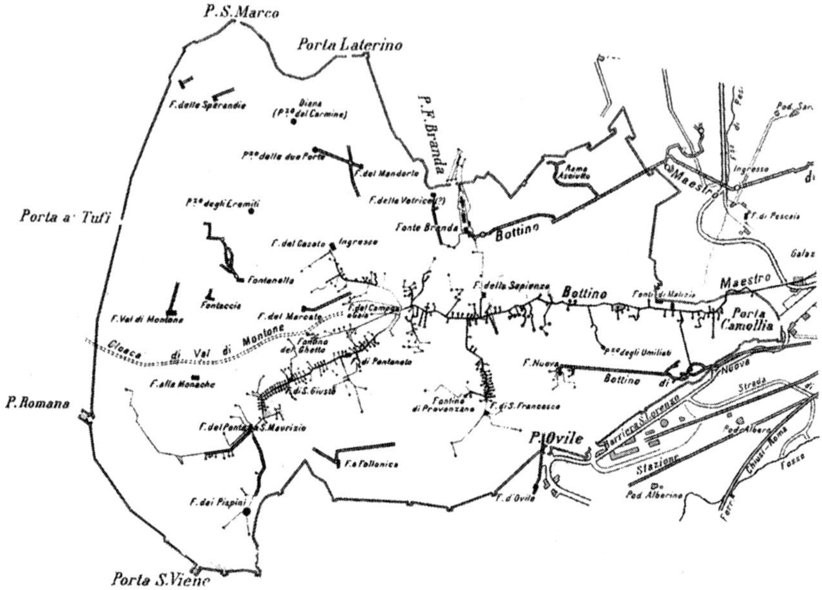

Figure 3.4 Diagram of underground Siena around 1900 showing the Bottino Maestro entering the city from the north (top of image) with branches leading to dependent fountains and cisterns (round dots). Also illustrated are the "spring bank" fountains with their own bottini, which are not connected to the main aqueduct. In the upper part of the image is the aqueduct feeding the Fonte Branda complex.

Although dividing the physical description of water supply systems into components such as "source" and "conduit" is often possible, the peculiar facts of Siena's older bottini-based water supply system will not permit drawing such clear boundaries. In Siena the bottini function as both sources and conduits of underground water. To avoid confusing the Sienese conduits with the type one finds throughout the Roman Empire, the older Branda type of tunnels will be designated by the local term *bottino.* The word is of uncertain etymology but may derive from *botte,* the word for barrel, and by extension, the sort of roof vaulting in the tunnels. The word bottino and its Latinized version, *buctinus,* have been used in Siena since at least the thirteenth century to denote the city's underground aqueducts. In this study, the newer, Gaia type of tunnel will be called an aqueduct. All the water supplied to Siena's fountains flows through subterranean aqueducts, of one type or the other, carved out of the soft, calciferous, sedimentary rock locally referred to as tufa.[22]

The bottino is merely an extension of the spring bank, a shaft driven horizontally through the aquifer. The aqueduct that feeds the Fonte Gaia functions more like a Roman aqueduct or Persian qanat, in that it acts solely as a conduit. Most of all it resembles a qanat, not only in principle but also in construction. The newer type of aqueduct differs from its Roman analog only in so far as it usually consists of an unlined tunnel carved directly through the tufa, whereas the Roman aqueducts were almost always lined. Qanats are only lined when they pass through a geologic formation—such as sand or gravel—that needs structural reinforcement. In his introduction to the subject of the Sienese aqueducts, Bargagli-Petrucci says, "In the bottini, more easily than the fountains, we can recognize a Roman principle." Bargagli-Petrucci makes the distinction between the two types in his description of the aqueducts and describes the different surveying methods used in their construction. Yet he fails to acknowledge adequately the critical departure from Roman principles the Sienese bottino and aqueduct represent.[23]

Although closer examination of the Sienese bottini shows them to have been built on a principle that might be Roman, the methods of construction are far from Roman. The function of Roman aqueducts was to convey water directly from a chosen source, not to collect it along the way. The surfaces of Roman aqueducts were designed to be non-porous. They were carefully plastered to keep pure water in and contaminants out. Even when they were raised high on arches to cross a valley, Roman aqueducts were enclosed on top to avoid contamination from wind-borne dust or animals. Not even rainwater could enter. The notion of an "aqueduct" that was porous, and thus vulnerable to contamination, would have been foreign to Roman engineers.

The last way in which the Sienese bottini and aqueducts differ from their Roman counterparts is in the "style" of engineering to which their respective builders subscribed. Lynn White, Jr., characterized Roman engineering as clumsy. To this we could add that geometric regularity often typified Roman engineering. Whether the dome of the Pantheon, the regular arches of the aqueduct piers, or the rectilinear street plans in Roman colonies, Roman projects tended to reflect geometric regularity wherever possible. Their regularity, together with their gigantic scale, produced a dominating effect on the landscape. The engineering of Siena's aqueducts is almost self-effacing by comparison. The patterns are organic, reflecting the contours of the streets and the landscape. Roman geometry is rectilinear, typified by the grid formed by the water mains and streets of Pompeii. If it were not for Siena's monumental fountain structures, there would be no visible evidence of the aqueducts' existence. Any Roman principle extending beyond a functional resemblance shared by the two systems is well concealed.[24]

The great difference between Sienese and Roman styles leads to speculation concerning its cause. The prime factors appear to be geology, scale, existing street patterns, and available resources. The Sienese system evolved over several centuries into a form that would best exploit the hydrological resources of its site. In contrast, Roman engineers, especially outside of Rome, were working in the context of an expanding realm. They sought a solution that could be easily replicated in dozens of new colonies. They devised straightforward, easy-to-implement plans. Hence Roman design tended toward regularity and standardization. It was less site-specific and less organic in nature. Finding the money to pay for expansion of the bottini system was difficult enough without adding the cost of making the unseen portions of the system as geometric and as regular as Roman work. Some of the irregularity of the Sienese forms can probably be attributed to the stop-and-go nature of the city's public works projects; financial difficulties or wars could interrupt projects, making it difficult to ensure a continuity of style through potential changes of personnel. The organic layout of the Sienese system resembled the pattern of streets in Siena, as it frequently paralleled them underground. In many Roman cities, especially new ones, the streets and water mains followed a grid. The Sienese layout conformed to the norms already set for streets.

The original type of bottino functions in a distinctly different manner from Roman aqueducts. I was quite surprised, upon my first visit to a bottino of the Branda type behind the Fontanella fountain, to realize that the entire length of the tunnel served as the collection device. The tunnel did not lead to a source; it meandered through the source—the water-bearing tufa into which it was cut. Because mineral-laden water is always seeping in from above, the walls and ceilings require constant arduous maintenance to keep them from becoming clogged with calcium deposits known as sinter. To increase the water supply, one had to increase the surface area of the bottino that cut through the aquifer. Usually this was done by simply lengthening it, but several constraints often prevented such a straightforward approach. First, the irregular surface topography caused the water to penetrate the subsurface in different quantities at different points. The burst of building activity that accompanied Siena's demographic explosion in the thirteenth century added further complications to the natural variations in terrain. Paved streets and gutters diverted the water from its normal point of entry as did compacting or plowing the earth. The results of the attempt to cope with these hydrologic and topographic irregularities can be seen in the shape of the bottino of the Fontanella, beneath the monastery of St. Agostino. Without the space to dig in a line, the excavators simply dug a loop, connecting the tunnel back with itself.[25]

When the bottini had been exploited to their maximum potential, the still-growing city had to find new sources of water or cease to expand. To do so required exploiting sources uphill from the city and bringing them in a secure conduit inside the walls. Ideally, the water would arrive at a more convenient location than the existing fountains. It appears the city recognized the finite nature of the water beneath it, because in 1334 it ordered that any new system could not disturb the existing system.[26]

From this matrix of competing demands Sienese artisan-engineers devised a new type of aqueduct in the fourteenth century. At first glance it appears only to be a longer version of the bottino. Upon examination, however, the new type resembles a qanat. The aqueduct called the *Bottino Maestro* (main aqueduct) of the Fonte Gaia is the best example of the new type. The walls of the new type of aqueduct do not constantly drip water as those of the bottini do. In fact, the intrusion of water through the walls is seen as presenting a risk of contamination from the streets above. To increase the supply of water to a fountain supplied by aqueducts, the tunnel needs to be extended to a new source. The result is that the aqueducts extend beyond the city in a pattern of branches, each to a different spring.

The distinction between the two types of aqueducts has become somewhat muddied over time, but it was not lost on those who developed the new system. A singular incident will illustrate this point. In the year 1356 Domenico and Giacomo dell' Acqua disturbed the governors of the city by pounding on the door of the Public Palace. They said that they had come, in the name of their persecuted family, to ask the city for money in the memory of their father, Giovanni, and grandfather, Giacomo, "who were *the real inventors of the entire system of bottini* and of fountains." For the sake of the city's honor, they argued, the £150 per annum granted to their father was not enough because he did not live long enough to enjoy it. Domenico and Giacomo believed that their father and grandfather, in building the Bottino Maestro, had devised a new method of conveying water, and their belief had a basis in fact. The aqueduct their ancestors built in the fourteenth century was based on a principle previously untried in Siena—conveying water to the fountains from a point beyond the city's walls through a non-porous channel resembling a qanat.[27]

The contract the city originally awarded to their grandfather gave the dimensions for the cross section of the Bottino Maestro: 0.9 meters wide by 1.8 meters high. The size, as with any excavation project, was determined by the minimum space required to accommodate an adult worker. The water itself fills only a small channel at the bottom, generally square in cross section, occupying between a third and a quarter of the width of the tunnel. To construct a bottino the excavators needed only to follow the barrier that the impermeable layer of clay formed beneath the aquifer. This required skills of observation but no special surveying equipment. If the construction crews

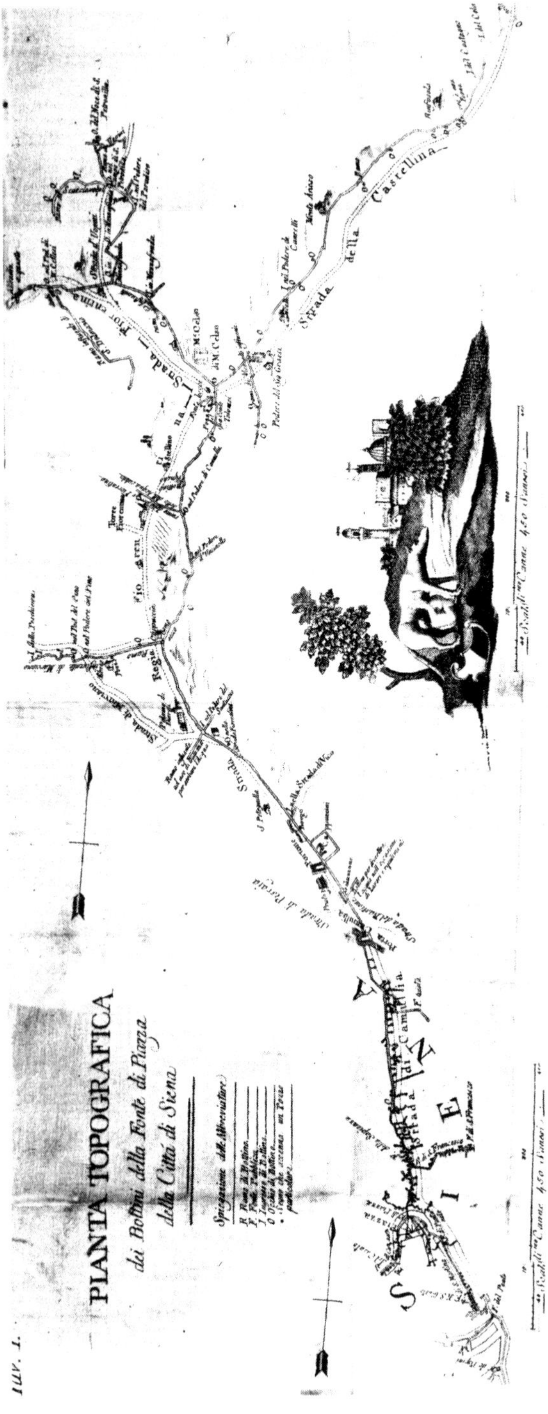

Figure 3.5 System of branching aqueducts extending into the Chianti foothills, drawn in 1799. From Battini, *Ricerche.*

dug too deep they would encounter clay and notice that the water was collecting instead of flowing downhill toward the fountain.[28]

The new type of aqueduct required much more careful surveying because it had to bring water a much greater distance than before, from various sources outside the walls to fountains in the city, like qanats or Roman aqueducts. The engineering requirements were stringent: to maintain a minimum gradient while following the most direct route possible along the contours of the hilly terrain to minimize costs. As with any other aqueduct construction, the gradient was the shallowest possible: Bargagli-Petrucci gives figures ranging from 0.25% to 0.9%, which neatly straddle the figure of about 0.5%—the average minimum slope for Roman aqueducts. The similarity in figures more likely reflects the laws of hydraulics than any direct transmission of Roman techniques or principles.[29]

Because the horizontal alignment was critical, the aqueduct is punctuated by frequent vertical air shafts, called *spiragli* or *smiragli.* These shafts are built like well shafts in that they are brick-lined and extend above the surface of the ground for about a meter. The extensions above ground prevent debris or surface water from accidentally entering the aqueducts. Besides helping the master builders maintain a constant incline, the shafts allowed them to choose the most direct route through the hilly countryside. In addition to providing a means for aligning the route, spiragli allowed fresh air to reach the workers below and provided a route through which the spoils could be evacuated, as well as easier access for maintenance workers after the system was built. Large stone covers installed over the openings prevent contamination of the water. The air shafts are essentially identical in form and function to the air shafts used in qanat construction.[30]

In both bottini and aqueducts the city built settling pools, called *serbatoio, gallazze,* and *gallazzone* (to denote the largest ones). By slowing the rate at which the water flows, the settling pools cause most suspended solids to precipitate, purifying the water before it flows into the fountains. Settling pools have been used to clarify water at least since Roman times. Vitruvius describes a *castellum* (reservoir) in his treatise, *On Architecture.* Of the benefits of settling, he says, "when the sediment has a place to settle in, the water will be more limpid and will keep a flavor unaccompanied by smell." Frontinus says of the waters upon reaching the reservoirs, "Here they take their breath, as it were, after racing through their channels and here they deposit their sediment." The use of settling pools was not limited to ancient Rome. Prior Wilbert, for example, built five of them in 1153 in the system that supplied the Benedictine priory of Christ Church, Canterbury, with water. So it is not surprising that Siena built a large settling pool immediately outside the city, under the Prato di Camollia, in 1438 at the cost of £942 6s. 4d. Minutes of the city council justify the expense with the following words: "so that the water should be clear when it arrives at the fountain in the Campo." From the settling pool the water flows into the aqueduct beneath the city. The water's motion through the channels and settling pools is so slow as to be imperceptible. Through it one can see the bottoms of the pools, encrusted with several centimeters of calcites fanning out in a delta from the point of entry.[31]

Bargagli-Petrucci contends that the function of the large pool was not only to purify the water, but also to maintain a reserve to offset the diminished flow during the driest months of late summer. It is more likely that its load-leveling effect was restricted to compensating for daily disparities between supply and demand. R.J. Forbes ascribes a somewhat more sophisticated method of "balancing the demands of the groups of consumers during every period of the day" by building a triple receptacle below the castellum from which pipes draw water. The hierarchal arrangement of three receptacles assures that public fountains are given priority over baths and private users in the event of a shortage. Smaller reservoirs, like the one behind the Fontanella, could be used both to clarify the water and to level diurnal loads. By allowing impurities to settle out in an easily accessible reservoir, the cost of maintaining the channel along the length of the bottini and the fountain basins would have been reduced. From the pools, the water enters the fountain basins through openings along the rear walls.[32]

The two aqueduct systems together comprise around twenty-six kilometers of underground conduits. Each of the older fountains is fed by its own bottino. The longest and oldest of these is the main aqueduct of the Fonte Branda, which is about four kilometers long. Each of the major fountains was, at one time, furnished with water from its own bottino. In decreasing length the rest are: Peschaia at eight hundred meters; Nuova at seven hundred fifty meters; Becci at five hundred meters; Fontanella and Follonica at two hundred meters; Ovile at two hundred meters; and finally the Fonte del Mercato, Val di Montone, Mandorlo, Sperandie, and Due Porte, each with a bottino less than two hundred meters in length.[33]

The Bottino Maestro feeds a single, integrated system superimposed on the earlier one of separate bottini feeding individual fountains. Originally conceived as a means of supplying the Fonte Gaia, the aqueduct soon came to supply the Fonte del Casato, the Fonte San Maurizio, the Fonte dei Pispini, and hundreds of private wells. From the sources north of the city to its many branches within the walls, its length totaled about eighteen kilometers. It brought water to elevations and districts where people had previously relied upon cisterns, wells, and water carriers.[34]

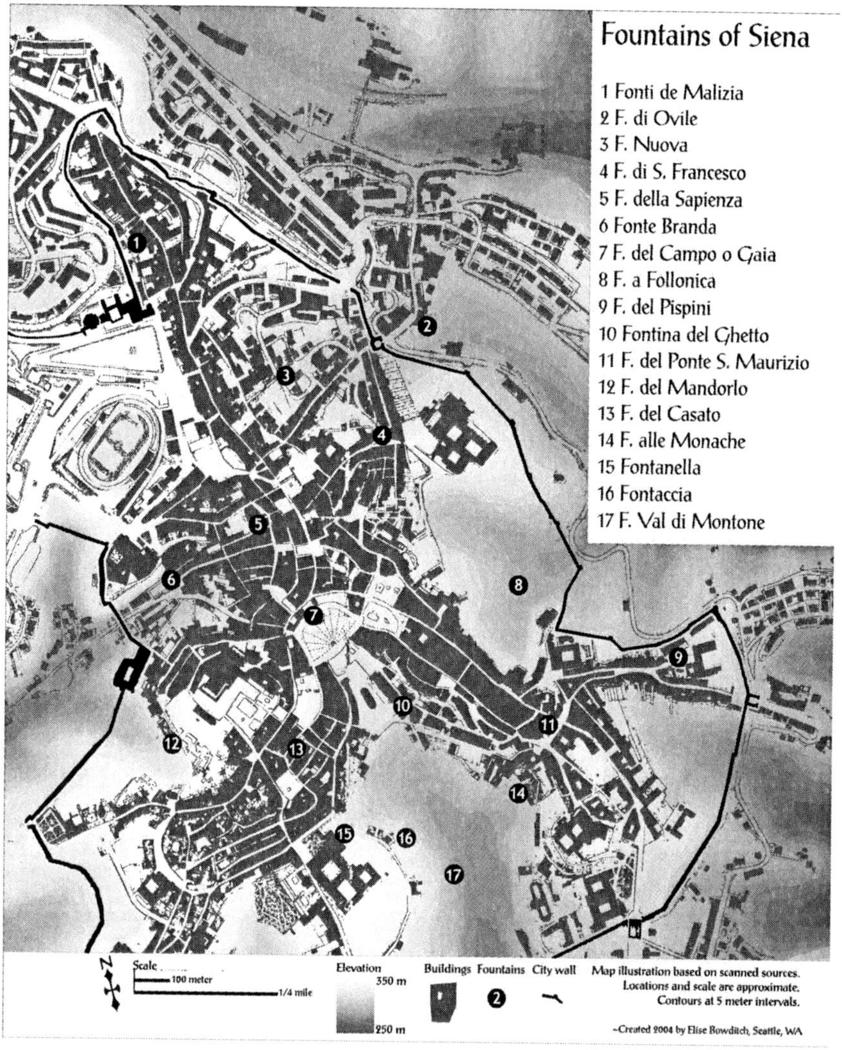

Figure 3.6 Map of Fountains of Siena. By Elise Bowditch, Seattle, WA. Lighter areas denote higher elevations.

From a discussion of Siena's system of underground conduits it is now necessary to turn to the fountains themselves. The design and location of the older system of bottini-fed fountains will be described, focusing on the example of the Fonte Branda. Then the ways by which the Fonte Gaia and the aqueduct that fed it fundamentally re-ordered the pattern of water distribution in Siena will be explored.

FONTE BRANDA

Siena's oldest and grandest fountain is the Fonte Branda, a thirteenth-century Gothic brick structure, with three ogival arches along its west and south facades, built on the site of a fountain that first appears in the records in 1081. It is built into a hillside in the valley that divides the cathedral from the church of San Domenico along the road leading from the Branda Gate to the center of town. It measures approximately nine meters high by twenty-four meters across the front (southern) face and nine meters deep. Each of the three arches is echoed by an inner, squatter ogival arch, recessed about half a meter behind the outer arch. The inner arch is much deeper than the outer, about two and a half meters thick. Together, the combined overhang of three meters provides the fountain's users shelter against both the winter rain and summer sun.[35]

Figure 3.7 Fonte Branda around 1900, with *lavatoi* (laundry basins) to left and laundry bleaching in the sun above. From Bargalgli-Petrucci, *Fonti di Siena.*

Whereas most of the fountain is built of brick, the lower portions of the piers and arches are constructed of cut grey stones, with narrow mortar joints. The inner arches are built of this stone, as are the lower portions of the piers of the entire facade, up to a height of four meters. Next to both sides of each arch is a worn head and torso of a stone lion sculpted in a Romanesque style. The four lions could have once been the water spouts mounted inside the fountain, from which the water originally entered the large covered basin. A document from 1246 records the installation of lions at the Fonte Branda. The city reimbursed 55s. to Master Iohanni for the cost of a new lion. The use of the word "new" suggests this was not the first lion to be installed. Water spouting from the mouths of animals was a frequent sculptural motif that can be traced back to the classical and even earlier periods. The choice of building materials, the scale of the structure, and the fine decorative details suggest the importance of the fountain to the city that built it.[36]

The fountain superstructure covers and protects a single large basin of water measuring about seven meters wide by twenty meters long by one meter deep. Several layers of plaster, now darkened with age, prevent discerning whether the basin was once partitioned into two or more sections. Such a

Figure 3.8 Fonte Branda facade around 1900. Note the marble lions' heads above; the brick arches resting on stone piers, and the city's *Balzana* affixed to the central arch. The facade is almost twenty-five meters wide.

division is hinted at by the multiple lions' heads at the ends of several parallel water supply outlets and would have facilitated draining and cleaning the pools without having to close down the entire facility at once. The insides of the barrel-vaulted ceilings are plastered as well, concealing the material of which they are constructed. The likely material is brick. The barrel-vaulted interior ceiling seems to be peculiar to the Fonte Branda. It may be a Romanesque element, like the lion heads, that survived the Gothic-style reconstruction in 1193. Little is known about the physical layout of the original fountain, though perhaps archaeological analysis would reveal parts of it within the structure of the current fountain.[37]

The capacity of the basin, approximately 140 cubic meters of water, extends the load-leveling function of the reservoir pools. The basin allowed water collected at night to be used the following day, minimizing waste. Four of Siena's five other principal fountains, the Fonte Ovile, the Fonte Nuova, the Fonte Fontanella, and the Fonte Peschaia share the same general pattern of the Fonte Branda, differing more in style than function. On the other hand, the Fonte Gaia, like its aqueduct, is built on a different principle and largely serves a different function than the ones built before it. Because the Fonte Branda is not only the most esteemed but also the best documented of the enclosed fountains, it will serve as the model and point of comparison for its type.

Figure 3.9 Interior of Fonte Branda, around 1900, looking west. The rear wall is on the right side. From Bargagli-Petrucci, *Fonti di Siena.*

Figure 3.10 Fonte Branda in section. From the front of the facade to the rear wall is about nine meters. From Strack, *Brick Architecture in Italy.*

The Fonte Branda is typical of several enclosed fountains connected to the original system of bottini that survive in Siena today. Large as they are, each fountain now constitutes only the partial remains of what were once large industrial fountain complexes. Each complex once covered four or five times more area than the structure that survives. Each complex consisted of at least three pools of water, each serving different purposes. The statutes and other archival references to the multiplicity of fountain-related activities, both sanctioned and forbidden, make little sense to a researcher today unless one realizes that the word for fountain, *fonte,* can mean two different things. Fonte can refer either to an entire complex of structures, pools, and connecting waterways serving many purposes, or it can refer to just the enclosed basin of potable water that is one part of that complex.

Siena's oldest fountains, like the springs that feed them, are distributed around the perimeter of the city, downhill from where most people lived. These majestic Gothic structures supplemented the city's wells and cisterns by providing water to those who needed a greater quantity than the two other sources could provide. It is likely that fountain water was generally of a higher quality as well, since it was always flowing.

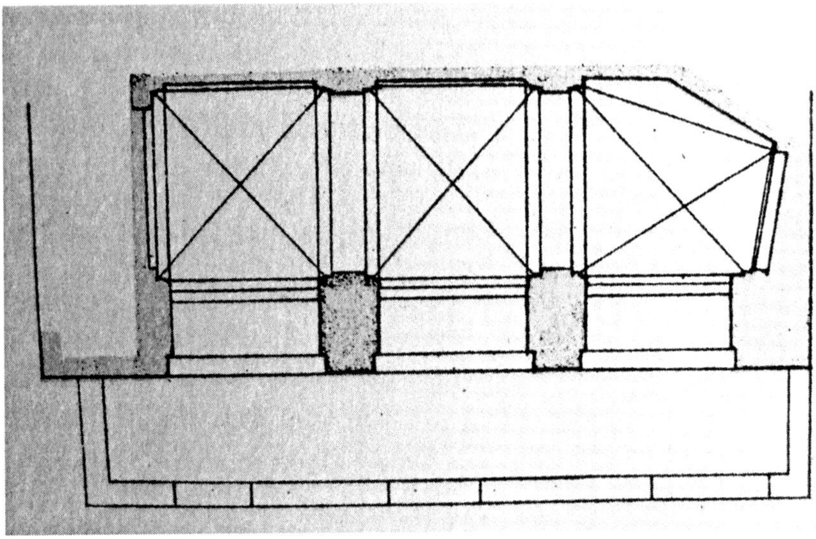

Figure 3.11 Fonte Branda Plan. Note the irregular shape of the bay on the right. A possible explanation is that it was the first built and that the fountain's orientation was shifted when it was expanded to include the two bays to the left. From Strack, *Brick Architecture in Italy.*

The uses to which water was put, when analyzed, clearly exhibit a hierarchical arrangement. The fountain builders physically embodied the hierarchy of water use in the structure of the fountains themselves. Because the fountains consisted of a series of connected basins filled from the rear, they did not have spigots under which one could place a vessel. To obtain drinking water, people had to dip pitchers into the front of the fountain basin. Therefore, strict observation of the hierarchy of uses was essential to maintaining the degree of purity needed for each use. By building the hierarchy into the structure itself, the city simplified enforcement of its regulations.

Two descriptions of the Fonte Branda, the first published around 1556 and the second in 1625, provide the earliest detailed narrative portraits of the many functions the Branda complex performed. Though the descriptions date from as much as two centuries after the period covered by this study, they remain useful. The ease with which the earlier legislation regulating the fountains can be mapped upon each of the described components allows us to construct a schematic image of the facilities at the Fonte Branda. Since the Fonte Branda is the paradigmatic industrial fountain in Siena, we can apply this schema to the other major fountains.[38] The first description says:

> Fonte Branda has so much water that a huge basin amply serves the city as well as filling a trough for the horses, laundry basins, and pools for bathing the horses; then the water flows underground and works more mills and wells and serves the wool guild, the leather guild, and other uses.[39]

Seventy-five years later, Sienese historian Giugurta Tommasi wrote a slightly more detailed account of the same fountain:

> Branda is the most copious of all the fountains. After the big horse trough, the water flows into the laundry basin to whiten the linens, and then making a pool that serves first to bathe the horses and then to wash the intestines of the carcasses—which are carefully butchered for the use of the whole city. The water then descends outside the walls of the city, where it powers a mill. Afterwards it forms pools for cleaning wool and cloth, and for curing the hides in a lime bath. Finally, before it joins the Tressa, it powers nine other mills with much benefit to the city.[40]

All these processes, including the supply of drinking, washing, and industrial water, and the operation of ten mills, took place within a narrow valley about 1500 meters long.

Figure 3.12 Fonte Nuova around 1900. Here the women pose for the camera. Behind them, barely visible, is the *lavatoio.* From Bargagli-Petrucci, *Fonti di Siena.*

Drinking water dominated the hierarchy of all the uses to which water was put. The importance of drinking water was so obvious as to be implicitly placed above all others. Tommasi felt no need even to mention drinking water in his description. Perhaps those citizens of his elite background never procured drinking water directly from a fountain themselves. It is likely that Tommasi did not have to think much about drinking water because he probably lived in one of the hundreds of houses supplied directly by the Bottino Maestro. Indeed, in his account of a journey through Italy forty-five years earlier, Michel de Montaigne reported that "Everywhere fountains are plentiful, but it is said that individual persons draw water privately from the conduits for their particular use." Nonetheless, the hierarchical arrangement Tommasi describes, if observed, would protect drinking water from contamination from other uses.[41]

The medieval statutes confirm the second use Tommasi described, watering the horses. For instance, the *Costituto 1309–1310* contains a provision requiring that the *Operaio del Duomo* (Master of the Cathedral Works) isolate water from inferior, "spongy veins," divert it from the fountain intake, and use it to feed the horse troughs directly, where it presumably joined the overflow of drinking water.[42]

For those not fortunate enough to enjoy a direct connection to the aqueduct, there were a few water carriers in Siena who could deliver potable water to where it was needed. In his study of thirteenth-century Siena, Daniel Waley found the water bearers (all of whom were women at the time) to be the lowest paid workers in the city, earning less than donkey-drivers or messengers. The reason for the relative lack of water carriers in Siena, compared for example to Paris, remains unknown. The low wages suggest that even the small number of water carriers in Siena constituted an oversupply of laborers. The low barriers to entry and the ready availability of female laborers—reflected in their low wages for excavation work in the bottini—could also have depressed wages. In Paris water carriers numbered in the hundreds and formed their own guild, which could have kept wages from becoming as depressed as they were in Siena. There is no evidence of a guild of water carriers in Siena. After the completion of the Fonte Gaia aqueduct, most of those who could afford to have home delivery of water already benefitted from a direct connection to the Bottino Maestro and were the likeliest households to employ domestic servants or to possess slaves. Perhaps the need for numerous water bearers was peculiar to cities sited along rivers where domestic well water was likely to be contaminated by seepage and runoff. The availability of potable domestic water could have precluded the need for water carriers.[43]

Figure 3.13 Fonte Peschaia around 1900. To the left are the remnants of the other basins. From Bargagli-Petrucci, *Fonti di Siena.*

The third level in the hierarchy was laundry. How the laundry basins were used, or precisely by whom, is largely unknown. The little that is known about laundry must be inferred from the written record, a topic that will be discussed in the next chapter. Below the laundry were pools for horses to bathe in, presumably to refresh and to clean them after riding. The following chapter will discuss these uses and their significance in greater detail.

Despite the huge investment of human and fiscal resources Siena made in its physical infrastructure, the volume of water always fell short of what the Sienese people desired. The only measurements of the quantity of water available in Siena are from the end of the eighteenth and the middle of the nineteenth centuries, and these are only for the Fonte Branda. The first figure is of 935 cubic meters per twenty-four hours. The second set of figures consists of two measurements, from September 1866 and November 1868, taken at the end of the dry season during a period of extraordinary drought that lasted several years. In 1866, the author of the latter study, Vincenzo Cambi, in consultation with a Professor Campani from Siena's university, measured 651 cubic meters and, in 1868, 632 per twenty-four hours. Cambi and Campani believed that the drought fully explained the discrepancy between Battini's higher figures and their own. Although it is difficult and dangerous to infer much from measurements taken from a single fountain several centuries after the period of study, these measurements could offer at least an order of magnitude for the minimum amount of water available in Siena. By using the maximum plausible population estimate of fifty thousand people in 1328, and the minimum water figure from 1868 of 632,000 liters per day, we get an approximate figure of perhaps twelve liters per capita per day for domestic, public hygiene, and industrial uses combined. Insofar as it is likely that the Fonte Branda's output constituted between one quarter and one half of the city's water supply, exclusive of rainwater collected by cisterns, the figure could be as high as four times that, which would be roughly equal to the United Nations World Commission on Water for the 21st Century's recommendation of a minimum availability of ten gallons per day. The daily quota could be higher still if the population is reduced (as it was after 1348) or if rainwater is included. For comparison, per capita usage in the United States was about seventy gallons per day in 1999.[44]

The figures for some Roman systems will make these numbers a little less abstract. A. Trevor Hodge offers the following estimates: Rome's smallest aqueduct, the Alsietina, supplied less than two percent of Rome's water; it discharged 15,680 cubic meters per day. Lyon's smallest aqueduct, the Mont d'Or, discharged ten thousand cubic meters per day, about an eighth of the city's total. Arles's aqueduct discharged eight thousand cubic meters per day; Strasbourg's 2,160; Geneva's 8,640. Segovia's aqueduct, the smallest for which Hodge offers a figure, discharged 1,728, more than twice the nineteenth-century figures for the Fonte Branda. The Sienese water supply was thus a trickle compared to typical Roman aqueducts in other cities. The chronic inadequacy of Siena's water supply and the inconvenient location of its most abundant sources, such as the Fonte Branda, led the city to take new and more vigorous measures to augment its water supply.[45]

Estimates vary as to when the city was growing the fastest, but the modern consensus is that the population reached its peak in the half century preceding the plague of 1348. The proposals to draw water from the Merse and the Staggia rivers have already been discussed, but they stem from this period. A sign of how desperate the city was to find more water can be seen in its allocation of funds to find the Diana, the lost river that the Sienese believed ran deep beneath their streets. If they could only recover this inexhaustible source again, their water worries would be over. Unable to locate the mythical stream, they continued expensive projects to tap into known springs and streams north of the city. Although each step taken might seem incremental when considered in isolation, the combined result was a small revolution in the quantity and distribution of water resources in the city.[46]

FONTE GAIA

By any measure, the Fonte Gaia, which reached its present form after nearly a century of work in 1419, broke with the prior patterns of designing, building, and supplying fountains with water. With fewer technical factors constraining its design, its exceptional nature is evident above ground as well as below. Its origins lay in the late thirteenth-century creation of the Piazza del Campo and the shift of the city's political, social, and commercial center to this awe-inspiring space. Partly carved from a hillside and partly built upon a filled-in ravine, the gently sloping clamshell-shaped piazza is an expanse of public space many times bigger than any of the squares built before it, including the piazza in front of the Cathedral. Today, by retaining the gentle contours of the land, its organic shape continues to astonish visitors and comfort residents. The citizens of Siena use the natural amphitheater every day of the year, from early in the morning when elderly citizens can maneuver without being jostled by crowds, until late at night when small bands of students gather on the pavement. One of the few aspects of Sienese life that seems to have remained a constant over the centuries since the Middle Ages is the centrality of the Campo to civic life. Thus, it is not surprising that the city found this great space to be worthy of an exceptional fountain.

The Campo's pavement is divided into nine wedges, once again echoing the multiples of three that underlie many patterns in Siena. Its southern end is supported by a huge amount of fill. Initially the Campo looked out over the hills beyond the city, with only the mint flanking the southwest edge of the vista. Soon the Public Palace abutted it. Together with the mint and customs house, the Public Palace enclosed the lower edge of the piazza. By the middle of the fourteenth century, the Campo was completely enclosed. The only additions since then are the Loggia della Mercanza, the Piccolomini Palace, and the small shrine in front of the Public Palace commemorating the Black Death.[47] To maintain a sense of organic harmony, the oldest known building code required the private palaces on the perimeter to employ the same triple-arched Gothic windows that decorated the facade of the Public Palace.[48] The naturalistic harmony of the Campo stemmed from the unprecedented imposition of design ideals upon some of the city's most powerful private citizens.[49] Not a gift of nature but a work of supreme artifice, the Campo transformed the city when it was built and remains Siena's most distinguishing feature seven-hundred years later.[50] Thus its single flaw, the lack of a public fountain, was probably all the more galling to the creators of this almost perfect space. For various reasons the usual solution, to place a fountain at the site of a natural spring, would not work. The Campo was at a higher elevation than any existing spring bank within the city. Any spring that may have once been in the Campo's vicinity would have had its flow cut off by the construction and pavement of the city center. Most of the cisterns along its edge were in private hands, and their combined capacity was much too small for the volume of people who would need to use cisterns. The only solution was to bring water from a higher elevation beyond the city walls through an aqueduct.

With this fourteenth-century commitment to build a new aqueduct came a new freedom. The fountain did not have to be built into a hillside, so it did not need to be covered. For the first time, the presence of the fountain would not determine land use of its immediate surroundings as the various industrial pools usually did. The piazza itself determined the nature of the surrounding district: a vast public space flanked by stunning palaces and public buildings. It would be a market, a meeting place, and an open-air theater for civic and religious spectacles.

Performing new and different functions in the life of the city, the Gaia fountain would not create the nucleus of an industrial district. It did not need to be—and probably the planners never wanted it to be—surrounded by a complex of five or six basins and troughs filled with pieces of leather, woolen cloths, and dirty laundry. The early fifteenth century thus opened new stylistic choices to the designer of fountains. An architect could cling to the Gothic or could borrow from classical forms, only recently again in vogue. In 1408, the city chose architect Jacopo della Quercia, who fully exploited these new possibilities in his design. Before Quercia completed his work, at least one other fountain stood in the Campo.

Other than knowing that an ancient marble Venus adorned it from 1325 until 1357, we have only the vaguest clues about the first Fonte Gaia's appearance. Insofar as it was not built into a spring bank, we can be pretty certain that it did not have the towering ogival arches of the city's other fountains. In the fifteenth century, Lorenzo Ghiberti (1378–1455), writing toward the end of his life in his *Commentaries,* mentioned that the earlier fountain was briefly adorned by a marble statue of Venus atop a dolphin. Ghiberti attributed the sculpture to the fourth-century Greek sculptor Lysippus on the basis of a sketch by Ambrogio Lorenzetti that the Carthusian friar Jacopino di Torchio showed him when Ghiberti visited Siena in 1416. Insofar as Lysippus is only known to have worked in bronze, it is more likely that the marble statue was a Roman copy. According to Ghiberti, a council of citizens assembled during the new Sienese government of the Twelve and determined that the erotic, pagan image had caused all manner of trouble for the city and therefore it should be taken down, smashed, and the pieces buried in Florentine territory. In answer to the citizens' outcry, the government resolved that the statue be removed and smashed, an action that is recorded in a document of 7 November 1357. Another possible reason for its removal could have been an attempt by the Twelve, who succeeded the Nine in 1355, to "dethrone" one of the icons of the Nine and to thereby consolidate their power by putting their stamp on the Campo and its fountain, which it will be remembered could be seen from the windows of the Public Palace.[51]

It took almost a century, from 1334 when Giacomo dei Vanni (dell' Acqua) began digging the aqueduct, until 1419 when Jacopo della Quercia (1374?-1438) finished his work, to connect an adequate supply of water to the fountain in the Campo. For that entire period there stood only a provisional fountain from which to dispense water. In December 1408 the city contracted with Jacopo della Quercia to design a fountain with an elaborate sculptural program, including a portrait of the Madonna and allegorical representations of the Virtues and man's origin befitting the city's most ambitious fountain and its position in the city's most spectacular public space. The contract obliged della Quercia to complete the fountain within three years of April 1409, but he did not finish it until ten years later in 1419. Thus, della Quercia's Fonte Gaia was at least the second fountain structure to occupy this site.[52]

The Fonte Gaia was unlike any other fountain for several reasons. First of all, it was supplied by a purpose-built channel of water extending far into the countryside north of the city, which allowed it to be placed on a piazza instead of being built into a spring bank. Second, it was completely open to the air. Third, the chosen location, the recently paved Campo, was the commercial and ceremonial center of the city, surrounded by some of the city's finest public and private buildings. The setting created unprecedented demands and opportunities for the fountain's designers. Perhaps it was an awareness of the gravity of the design choice, or the political instability, or the lack of money that delayed the selection of a designer until 1409. The choice of della Quercia resulted in the city's most visible and greatest work of art in the "Renaissance" style that was emerging in Florence at the time. Della Quercia designed the fountain as both a civic symbol and as a source of drinking water, but not as an industrial water supply. Instead of many pools, it had only one large one, enclosed on three sides by low, niche-filled white marble walls. In the niches stood various allegorical and biblical figures carved in high relief. Six wolves crouch along the sides and back of the basin's walls, each with water spouting from its mouth. Water could be obtained by dipping a container into the basin or by holding it under one of the streams of water coming from the wolves' mouths. The unprecedented form (at least in Siena), the new style, and the virtuosic execution instantly made the fountain a landmark in the city's cultural history.[53]

Della Quercia's fountain stood until the second half of the nineteenth century, by which time its poor condition had become an embarrassment to the city, which commissioned Tito Sarrocchi to make a copy. The surviving remains of the old fountain are now displayed in the Loggia of the Public Palace and are well-documented in Anne Hanson's monograph.[54]

In the fourteenth century, while the city was spending decades to work out the details of the fountain's appearance, it was also conducting the analogous process of seeking an adequate water supply underground. The city let contracts to various masters who hired dozens of day laborers, both men and women, who chipped away at the tufa with picks and shovels. Masons followed, reinforcing weaker sections of the tunnels and lining air shafts with brick. The masters in charge missed their deadlines (including the dell' Acqua family mentioned earlier), the city extended deadlines, and new promises were broken yet again. The city canceled contracts when only a trickle of water arrived, and it hired new masters. In 1366 the city suspended major work, except for cleaning, for a period of sixteen years, until 1382. The government appointed new committees to oversee the project. This administrative nightmare went on for centuries.[55]

A number of causes contribute to this disarray. First, before the laborers excavating the tunnels had even finished their work, various private parties had already set their eyes upon the aqueduct passing in front of their houses. Millers, struggling to make do with temperamental trickles of water in the southern districts of the city, pleaded for relief and a branch aqueduct of their own, which the city granted them, even before it had enough water for the public fountain. Second, beginning in 1355 with the fall of the Nine, Siena was never to have as long-lived and stable government again. The reader will recall that in the autumn of 1368 alone four revolutions occurred. Third, the technical task itself was unprecedented; the Bottino Maestro represented a new and untried technology in Siena that the engineers and builders figured out only as they went along.[56]

We can only speculate on how the city's citizens regarded this situation. As in other cities, the new aqueduct would not affect each person equally. For the abutters of Siena's main north-south artery, an aqueduct passing beneath the street in front of their houses promised a welcome amenity. It held out the potential to bring running water, where it had never been before, to the cisterns and wells in their courtyards. For dozens of skilled and unskilled workers, the new aqueduct would mean years of round-the-clock paid labor. But for those women living at the margins, supporting themselves by hauling water to the homes of the rich, the new aqueduct would likely spell ruin.

SIGNIFICANCE OF DESIGN

The hidden complexity of Siena's aqueducts is mirrored by the elaborate architectural and sculptural monuments that mark the system's urban outlets. These magnificent brick edifices, built with a spare design, grand proportions, and ogival arches characteristic of the Sienese Gothic style, served not only to protect the springs from cave-ins and contamination but also to make access to the water easier for all its consumers. The monumental scale and elaborate decorative programs of the fountains did more than facilitate access: they expressed water's significance to the people of Siena. Furthermore, whereas the grandeur of the fountains symbolized the extent of the city's investment in the water supply, the uniformity of their design expressed the city's centralized control of the system.

When the city built a new system to bring water to the Campo, introducing to the Sienese context an unprecedented type of aqueduct, it adorned the terminus with an equally unprecedented fountain, the Fonte Gaia. The use of white marble and a plan whereby the three walls remained open to the sky, punctuated by niches displaying allegorical figures, expressed the break with past styles and techniques. The city took great care to choose an appropriate designer, settling on Jacopo della Quercia. The choice of one of the era's finest sculptors demonstrates how strongly the city's pride and civic glory were tied to the water supply system and the fountains that displayed and distributed its bounty.

Except for individual wells predating written records, the entire system seems to have always been administered by the city and not by private entities. Maintaining roads, bridges, and fountains by public authority seems to fit a long-held pattern in the rest of northern Italy. Exceptions to this rule were few. Siena had relatively few public water carriers and few regulations governing the sale of water. Although the central government often delegated administrative functions to special committees or to neighborhoods, the supervision of water supply remained in the public sphere. The mechanisms by which the city governed the construction of the water supply system paralleled the means by which the city oversaw the erection of its physical defenses.[57]

Only through municipal administration could complex, expensive, and long-term projects be brought to completion and kept running through numerous changes of personnel and the rise and fall of entire governments. Only a central government was able to marshal the fiscal and expert human resources necessary to build and maintain such a system. Likewise, only the city government was in a position to recognize the need for maintenance and to act upon those needs in a timely manner.

Despite periods of neglect and destruction by war, the Sienese water supply system has enjoyed remarkable longevity—at least a thousand years of service to at least fifteen thousand people. Not only has the system endured, but also it survives in an incredible state of preservation. Most of the visible as well as the hidden components have survived intact since the Middle Ages. The introduction of a modern water supply system followed the same pattern as the two earlier phases of expansion: the new system supplemented the prior ones without destroying them. One of the unexpected results of providing running water to households was the consequent necessity of providing modern sewers to remove water and water-born waste. The provision of sewerage proved to be one of the more disruptive public works projects in this century, requiring the excavation of nearly every street in the city.

The city and its citizens built a total of four distinct water-supply systems over the course of a millennium: cisterns, wells, spring banks, and aqueducts. Each of the systems overlapped the scope of its predecessors. The number of different systems increased the variety of technical challenges faced by Sienese engineers. Furthermore, the redundancy of systems offered the city's water supply resiliency in the face of disruption. Another feature of the presence of multiple systems was the availability of water in several grades of quality and purity. The varying levels of purity made it likely that the patterns of distribution and use should evolve into a hierarchical arrangement. The ways in which the city ensured the best water was reserved for the highest uses, as Alberti recommended, will be the subject of the next chapter.

Chapter Four

Regulation of Water

This chapter will focus on the Sienese water supply system from 1250 to 1348, during which time the city reached its peak population of around fifty thousand people. During the same period its rulers promulgated twenty-six sets of statutes. Much of the population growth was not internal, but is attributed to immigration of people from the countryside. It may well be the very waves of immigration of rural newcomers necessitated the constant reiteration of regulations governing conduct of life in an unfamiliar, urban setting.[1]

The government of Siena took two parallel approaches to imposing order on the use of water. It wrote laws regulating the use of water, and it built fountain complexes that embodied the provisions of that legislation in their physical arrangements. The goal of each approach was the same: to impose a hierarchy of use whereby those uses of water with greater potential for contamination were kept downstream from the uses that required a supply of pure water. This strict hierarchy was necessary because water was used and re-used several times before it was allowed to leave the city. A secondary goal was to avoid conflict between users of this scarce resource by setting predictable allocations to each. A wealth of Sienese statutes addressing water use survives from the thirteenth through the sixteenth centuries.

Policy, in the sense of a generalized set of answers to an anticipated set of problems, evolved slowly and incrementally. The laws dealing with specific aspects of Sienese public life could be traced by comparing the sixty-five sets of statutes preserved in the State Archives of Siena. However, rather than following the evolution of policy itself, this chapter analyzes the statutes to explore how the city attempted to accommodate the often competing needs of users of water in the thirteenth and fourteenth centuries. It also examines specific uses of water, such as for industrial purposes, firefighting, and personal and domestic consumption by looking at statutes, account books, and the physical remains of the fountains and aqueducts that supplied them to shed light on day-to-day practice in Siena, comparing the prescriptive nature of the statutes with the reality of water use in the city. Finally, this chapter will assess the success with which the city met its goals.[2]

The Sienese people built their water supply system in response to a specific set of environmental, political, economic, and social constraints. To understand why the water supply system took the form it did it is necessary to keep in mind those siting constraints examined in earlier chapters, beginning with an explanation of why the city occupies several hilltops endowed with so little water. The selection of the site has its roots in the changing nature of travel between Northern Europe and Rome in the early Middle Ages and in the manner in which way-stations along this route, such as Siena, might have responded to these changes. During the thirteenth and fourteenth centuries, the over sixty fountains, large and small, supplied the city's water. Of these, at least half a dozen comprised multi-basin fountain complexes. Almost all of Siena's fountains were (and the surviving ones still are) fed by gravity through its own underground aqueduct. The exceptions were about eight fountains, all fed by the same longer aqueduct, called the Bottino Maestro, which was completed in 1343.[3]

The thirteenth and early fourteenth centuries proved to be a turning point in Sienese economic and political history. The city grew and prospered, but because its economy depended mostly upon a few slow-growing sectors such as agriculture, services such as banking and lodging, and regional commerce and travel, its relative power and wealth diminished as other cities in the region industrialized. The city's leaders struggled to reduce impediments to industrialization, but the physical limits of the city's hilltop location made it extremely difficult to compete with better endowed cities like Florence. As already noted, Siena slipped behind cities well provisioned with ample water, and running water to power mills provided an economic advantage that may have become particularly important after the Black Death of 1348. Although Siena's population grew and its economy expanded, it could not diversify growth as fast as other cities in the region and was ultimately eclipsed by Florence.[4]

As the city reached the peak of its population in the early fourteenth century, Siena's government became increasingly concerned with regulating the water supply to ensure that the city would continue to flourish. These concerns about water can be discerned by examining various laws the city enacted. For instance, an article among the Statuto 20, a set of Sienese statutes from 1309–1310, ordered that latrines and cesspools discharging through the city walls be removed "for the beauty of the city. . . ." Another regulation, from the Statuto dei Viarî, a collection of Sienese statutes governing public works redacted in 1290, ordered that a small gate in the city walls near the Porta Camollia be filled in because it prevented "the beauty of the city" from being seen. A third law ordered a section of the road to Florence repaired "for the honor and the utility of the City of Siena." From these isolated explanations of law we are tempted to infer that the city's beauty, utility, and honor were foremost in the minds of its legislators. However, such rhetorical formulas are not uncommon in Tuscan city-states. Usually the laws prescribed only the specific measures to be taken or practices to be avoided. The statutes included provisions for ensuring an adequate minimal supply of water for all users, preventing contamination of the water supply, disposing of polluted water properly, and confining noxious activities to specified industrial districts. The regulations applied to all users of water, both corporate and individual. The city published the regulations as the Statuti, folio-sized bound volumes kept at the *Palazzo Pubblico* (Public Palace), where copies were available for public consultation. Despite the frequent failure to state an explicit overriding philosophy, the consistency of the prescriptions does suggest that a regard for "the beauty of the city" did in fact motivate much of the regulatory process.[5]

Siena enforced its statutes by various means, including stationing custodians and guards at individual fountains and the use of secret informers. The city encouraged minor officials to press charges by returning half the fine collected to the complainant. Conversely, officials could be fined the full amount for not enforcing the law. Sanctions were generally monetary, but exceptional crimes, such as attempting to poison a fountain, could warrant capital punishment. To what extent the regulatory mechanism actually affected the use of water in Siena can only be known indirectly. The effects of regulation must have varied with the shifting fortunes of the city; but if the regulation of the use of water even approached the success the city obtained in regulating building and zoning, then it must have been remarkably effective, surpassing all efforts that followed until the nineteenth century. The Sienese statutes (and their enforcement) reveal a desire for a city that was clean and beautiful. They further disclose a faith in the rule of law and the belief that these ends could be accomplished through the regulation of citizens' day-to-day behavior.[6]

By looking behind the statutes one can discern aspects of the lawmakers' image of an ideal city. The vision of a clean and orderly city is corroborated by images in Ambrogio Lorenzetti's mural from the 1340s, "The Virtues of Good Government," in Siena's Public Palace. Lorenzetti depicts wide streets and broad piazze, which although heavily trafficked, are level, dry, and free of dung, refuse, and puddles. In one scene, elegantly dressed dancers protect their feet with colorful, form-fitting shoes—not mud-caked boots. Such images, in light of the contemporaneous statutes, suggest an urban ideal far removed from notions of medieval filth and disorder that persist in today's popular imagination. They reflect late twelfth and early thirteenth-century Sienese civic ideals that anticipate those that Leonardo Bruni would articulate one hundred years later in his book, *Panegyric to the City of Florence.* In the *Panegyric,* Bruni claims that "Florence has surpassed all other cities in its prudent site and its splendor, architecture, and cleanliness." He emphasizes Florence's superiority by noting that "every other city is so dirty." He concludes his comparison by saying that "filthy cities that may in other respects be very good can never be considered to be beautiful." Although Bruni's partisan intentions force the reader to distrust any comparisons he makes between sanitary conditions in Florence and other cities, it is safe to conclude that cleanliness is one of the criteria by which he judges cities beautiful. Seeking to criticize Siena, the Florentine humanist Leon Battista Alberti alleges foul smells pervade the city's streets. In sum, cleanliness can be most easily defined by what it is not: foul smells, filth strewn on streets, or contamination of water by mud, laundering cloth, or tanning leather.[7]

At least since the time when Vitruvius wrote his prescriptions for the architects of Rome, civic leaders have made an association between the appearance of their domains and the nature of their governance. After the Roman government lost control of the Empire ecclesiastical leaders filled the vacuum. As Italy re-urbanized, secular governors—it mattered little whether by election, imperial charter, or conquest—began again to share responsibilities for the maintenance and appearance of urban structures, roads, and public spaces. It would appear that the same sort of "civic spirit" that Richard Goldthwaite found emerging in the wake of Florence's prosperity animated the rulers of Siena as well.[8]

An examination of some of the provisions Sienese rulers made for the use and maintenance of the public water supply will reveal the concerns of the lawmakers. The Sienese communal statutes are preserved in sixty-five bound volumes redacted and collated from 1262 until 1545, ten years before Siena lost its independence. More than a third of the regulations date from the first three decades of Siena's longest-lived government, that of The Nine Guardians and Defenders of the Commune and the People of Siena, and thus form an important source for discerning that government's perspective. The Nine ruled Siena from 1285 until their overthrow in 1355. Each collection of statutes is divided into five *distinzioni* (chapters) according to the matter treated. Lodovico Zdekauer argues that the five-part division revealed the enduring influence of Roman law and that the Sienese statutes were consciously modeled on the sixth-century Code of Justinian. The third chapter governs the physical and administrative aspects of Siena's infrastructure, both within the city walls and in the surrounding territory. Zdekauer argues that devoting the third chapter to public works is both characteristic of and original to Siena. He notes that Florence and Pisa both put the criminal law into the third chapter, assigning public works to an appendix. It would appear that Siena consciously chose to place infrastructure in a more prominent position, a choice fully consistent with what we know about the government of the Nine. Most of the laws discussed here will be drawn from the third chapter of various sets of Sienese statutes. The general pattern seems to be one of attempting to conserve accepted norms during a period of population growth and physical expansion.[9]

The title of the first law in the third chapter in Statuto 20, "Regarding the walls, ditches, gates, moats, fountains, bridges, streets, and roads of the commune of Siena," offers a typical definition of the scope of the laws that follow. In addition, the first provision describes the complex apparatus for administering and enforcing the regulations. The *giudice sindaco* (Judge Syndic, the primary guardian of the laws) had to inspect the walls secretly, in the best way possible. The law required him to report any violations to the Lord Captain of the People, and the Lord Captain of the People had to accept his reports. Violators had to repair any damage or pay a £25 penalty. If the Judge Syndic failed to enforce the ordinance, the Four *Provvisores* of the Biccherna (Comptrollers of the Treasury) were to withhold the £25 penalty from the judge's salary. Several of the most salient features of the Sienese regulatory process can be seen in this first ordinance: the dependence on written statutes, the division of regulatory labor among a number of jurisdictions, the various layers of supervision and accountability, and the power to impose monetary sanctions upon the regulators themselves should they fail to enforce the law.[10]

The Sienese statutes are specific and concrete. Rather than attempting to achieve general applicability or to anticipate problems, each provision appears to have been written in response to a particular petition, deliberation, or public nuisance. Analyses of the Sienese statutes suggest an empirical means of setting policy. When examining the manuscript originals, one sees that many articles were crossed out and amended or revised in the margins, showing the steps by which laws changed as lawmakers' understanding of the city's needs changed. The consistency of provisions suggests that certain general ideas about the management of the city, for instance, the desire to allocate scarce resources, minimize costs, reduce conflict between groups of users, and to preserve the purity of water supply, underlay the specific provisions.[11]

Certain patterns immediately emerge from the documents. The first of these is the delegation of authority from paid officials such as the *Podestà* (an office held for six-month terms by a non-Sienese, akin to a modern city manager), Judge Syndic, and Four Comptrollers of the Treasury to ad hoc committees of volunteers. Although statutes were voted into law by a combination of the General Council and the executive body of the Nine, and ultimately enforced by the Podestà, small committees worked out the nuts and bolts of gathering the information needed to write laws and of carrying out the details of execution and enforcement. These committees were usually composed of three, six, or nine *boni homines* (good men), one third of whom came from each third of the city. Often one of the members was a notary who kept the committee's records. The city appointed these committees to find new sources of water, to design fountains, to oversee construction projects, and to supervise repair and maintenance programs.[12]

Such committees could have among their members certain experts, usually referred to in the statutes as *maestri* (masters). For instance, Statuto 20 required the Podestà to employ a good master stone mason and a laborer to

Figure 4.1 Fonte Follonica around 1900. Originally built on the same scale as the other major fountains with elaborate architectural decoration in the Sienese Gothic. Seventeenth-century views of the city indicate that for the past several hundred years it has remained mostly buried in the gardens inside the city walls, down hill from the church of San Francesco. From Bargagli-Petrucci, *Fonti di Siena.*

maintain the streets for the entire duration of his term, with the exception of feast days. If the city wished to retain a specific expert, the statute referred to him by name or by his title, as when a law from 1292 specified that the *Operaio del Duomo* (Master of the Cathedral Works) had to review a certain project. Another statute required that three wise and discreet men, one from each third, be selected to inspect the drainage ditches and moats once a month. A provision regarding the Fonte Peschaia required the Podestà to select two good men from each third of the city, who together with wise masters would improve the access road to the fountain.[13]

The exception to this pattern of appointing multiples of three came when matters were confined to a single district. In such cases the city only needed representatives of the affected third to oversee the work. For example, in 1309 the city selected four good men from the third of San Martino to inspect the Fonte Follonica (the now almost completely buried fountain that served the district) and to investigate the best place to build a laundry basin. The importance of representing each third or smaller division of the city, such as a *popolo* (an administrative unit roughly equivalent to a parish, of which there were about thirty-six in 1318), can be seen in the distribution of a committee to design a new fountain by the Porta Ovile. The article ordered the *Camarlèngho* (Chamberlain) and the Four Comptrollers of the Treasury "to select ten good and discreet men, that is two from the popolo of San Pietro of Ovile, and two from the popolo of the abbey of San Donato, and two from the popolo of San Andrea, to find the most useful and convenient place for the Commune of Siena to build a fountain inside the Ovile Gate." To address cases in which the city did not appoint a specific committee, one provision allowed any four or more men of a neighborhood to request that the city guard, clean, or repair the local fountain at the city's expense. Even when such exceptions are considered, the appointment of committees in which each third of the city was represented remained the norm.[14]

For the most sensitive sorts of work, such as estimating damages, appraising property, and solving accounting disputes, committees were composed of *viri religiosi* (religious men) or *uomini di Penitentia* (men of penitence), citizens who had voluntarily bound themselves to a higher standard of behavior. They did so either by making private vows or by formally joining one of the several tertiary orders affiliated with friaries in the city. Religious men could presumably be trusted to stay above mundane issues of money and not to take sides in a dispute. One of these crucial tasks was the inspection of the major roads in Siena's territory, including the via Francigena, the road from France to Rome. A statute ordered the city to elect two good, discreet, religious men to inspect the section of the via Francigena that was

Figure 4.2 Fonte Ovile around 1900. Note the women using the *lavatoio* (no longer extant). From Bargagli-Petrucci, *Fonti di Siena.*

under Sienese jurisdiction. Perhaps seeking to exploit both the technical expertise and moral stature of such men, the city frequently appointed religious men to oversee various aspects of its water supply system, as well.[15]

The range of legitimate uses and forbidden practices associated with the consumption of water, at least in public spaces, can be ascertained from reading the Sienese statutes. Duccio Balestracci and Gabriella Piccinni echo the distinctions Alberti made in his treatise *On the Art of Building* when they observe that the legitimate uses of water fell into three main categories: industrial applications, firefighting, and personal and domestic use. The first two categories of activities took place in the public sphere. They were more easily regulated by legislative means than domestic uses and thus better represented in the statutes.[16]

The hints about behavior behind closed doors are few and can only be partially inferred from the statutes. The regulations governed only those aspects of personal water usage that impinged upon public spaces or used public facilities. They included quenching one's thirst along the road, for which the city provided various wells and fountains; washing laundry; firefighting; and using the public baths constructed at the many geothermal springs in the Sienese countryside. This discussion of water use will describe its industrial applications in Siena, firefighting, and then its personal and domestic uses.[17]

Industrial uses of water can be divided into three categories: capturing the potential energy of water to produce motive power, employing water as a solvent or agent in chemical processes, and using it to flush away waste. The most highly regulated users inside the city walls were members of Siena's textile, butcher, and leather guilds, which exploited water's chemical properties as well as its ability to flush away waste products of their trades. Outside the city, where the greater quantity of water allowed the mills to use water's potential energy, the regulatory apparatus extended to mills located along the rivers miles from the city such as the Merse and the Farma.[18]

Unfortunately for hill towns like Siena, the application of power to manufacturing processes, especially in the textile industries, was ultimately to prove an important source of wealth and means of diversifying economies that had depended upon commercial and financial services. Beginning in the thirteenth century, Siena went to great lengths to increase the supply of water for both industrial and domestic users as the city's population grew. Its leaders consulted experts from inside the city and from abroad. Despite the failure of these efforts, Siena continued searching for new sources of water to satisfy industrial needs and its growing population, which was to peak around 1328. Insofar as none of these attempts ever overcame geographic limits of the site, the city seems to have turned its focus to regulating the little water to which it did have access.[19]

With the possible exception of a few mills using the overflow from some fountains (which were mostly near the walls anyway), all water-powered mills stood outside the city walls. The mills appearing most frequently in the documents ground grain and fulled wool. Because of the vital role mills played in keeping Siena supplied with flour, the city regulated every aspect of milling grain. Since bread played a central role in the nutrition of the vast majority of people, the failure to maintain a steady supply at an affordable price could result in starvation, the outbreak of epidemic disease, and a regime's worst fear, civil unrest. Milling flour and taxing millers were both profitable activities, providing an important source of income for the city. The economic importance of the fulling mills demanded similar attention to their operation. Indeed, an entire set of laws governed just the Woolworkers' Guild—a matter that will be discussed below.[20]

Underlying the scheme regulating mills was a system of riparian rights, which were intended to prevent conflict between competing users of the same water course. In a series of a dozen regulations addressing mills, the most general prohibited occupying any ditch in which water usually ran or impeding the flow of water such that the operation of any mill was threatened. A related type of statute attempted to limit the damage mills could do to other users. One statute ordered that a wall be constructed at Bagno Vignone so that the baths would not be emptied when grain was being ground, suggesting that the bathers and millers competed for the same limited supply of water. Another aimed to protect a fountain from being damaged by a mill pond, ordering that no ditch or mill pond be built within about 100 meters of the fountain in the Val di Montone. A third ordered anyone who built a mill race that crossed a road to build a bridge of stone, brick, oak, or [other] sound beams of optimum quality so that people and animals might easily pass over the new mill race. The bridge had to be at least two and a half meters wide; if the site made such breadth impossible, it had to be at least as wide as the existing road. As crucial as the interests of mill owners were, when riparian rights and land transport competed, the city considered keeping roads passable at least as important as regulating the use of the water.[21]

Most statutes, however, were intended to protect mills and mill ponds from external threats and from owners of nearby mills. The penalty for damaging mill ponds or races was steep: £25 per infraction in 1262. Such attention to mills and millers is not surprising considering that the city depended upon their uninterrupted service to prevent famine and to maintain a balance of trade. No doubt millers and their interests were well represented in the legislative apparatus. Provisions that protect such large investments seem appropriate in a city governed by banking and commercial interests.[22]

By the end of the thirteenth century, as the city's population reached its peak, mill owners and the city seem to have realized that there was a limit to how many mills could be built along any given section of a river. Reconstructions of mill locations in the Sienese countryside by Maria Elena Cortese do not suggest crowding. Terrain, especially the slope of the stream bed, is the primary limit of proximity. The first statutory limits on mill density—which appear to have applied to parts of the Sienese territory—date from 1262. It appears that when the city government perceived that the rivers had reached their capacity for providing power to mills, it capped the number of mills on all the rivers in its jurisdiction so that operation of the mills upstream would not be impeded by mill ponds backing up from those dams downstream, and vice versa.[23]

To ensure that no one could falsely "grandfather" a new mill by claiming it had always been there, another statute ordered the Podestà to make a written inventory of all mills along certain rivers. Such an inventory would also be quite useful to ensure that the tax rolls were up to date. The size of the fine in case of the Podestà's failure to comply, £100, suggests both the difficulty of enforcement and the importance of this aspect of managing the city. After the Podestà had completed the inventory, no one could erect a new mill or change an existing one in any way. The penalty for impeding the operation of another mill by modifying one's own was £100. The next statute required the violator to undo anything new. Although intended to avoid conflict in the process of distributing a scarce resource, such laws had the potential to impede technological changes, even ones that would improve efficiency, such as the introduction of a new water wheel. Although these regulations merely attempted to defend the riparian status quo against the overbuilding of mills and the possible self-destruction of the industry, they likely also had the effect of presenting considerable barriers of entry to prospective millers. Doing so would protect profits of established millers at the expense of consumers by potentially raising the price of mill complexes and thus the price of flour. On the other hand, by ensuring the health of the industry by governing its growth through mill construction, the government maintained a stable supply of flour and stability seems to have been an overriding concern of the Nine.[24]

Notable is the exception the above rubric made for mills owned or operated by the clergy or religious orders, whose consent was required before their mills could even be inventoried. Other regulations extended similar exemptions to the *Ospedale,* granting it permission to build mills without opposition from anyone. The freedom from legislative restrictions on innovation could have given religious institutions a technological advantage over other mill owners. This is especially likely in light of the ability of charitable institutions to accumulate an enormous amount of capital as well as technical expertise. The deference shown to the hospital was no doubt a result of the intimate and symbiotic relationship between the city and the hospital. The hospital was an autonomous lay institution that provided several kinds of social welfare in Siena. In return for its services, the city granted the hospital various privileges similar to those accorded religious institutions. Once again, the rulers seem to have designed a regulation in order to ensure greater stability.[25]

Another statute, in what seems to be an attempt to offset the limits on mill density, encouraged the construction of mills by extending the powers of eminent domain to builders of mills. It ordered that abutters sell their land to those who wished to build a mill. As usual, the statutes often exempted religious land owners. It is not surprising, considering the lack of unifying policy, that the statutes encouraging the construction of mills seem to contradict the regulations prohibiting the construction of new mills or altering extant ones. It was not unusual for the city to reverse itself periodically—a process that perhaps reflected shifts of power within the ruling oligarchy.[26]

The most widespread and problematic of all industries within Siena's walls, textile and leather production, employed water as a chemical agent and as a means to convey waste products away from the city. Processes that required power from water generally needed to be performed outside the city's walls, where sufficient water supply existed to power mills. A third major industry—intimately linked to tanning—butchering, posed similar environmental problems. Because meat was perishable, butchering was best performed physically close to the consumers, despite the mess it caused. Likewise, the best time to slaughter the animals was right before the people consumed the meat. Numerous regulations specified where one could and could not macerate linen, dye cloth, wash meat and skins, and tan leather. In addition to the city statutes, some of the *arti* (guilds) had their own statutes, regulatory processes, and enforcement mechanisms governing the activities of their members. Industrial users had legal access only to certain basins of specific fountains. Much of the textile district centered on the Fonte Branda and the pools below it, where the city took special care to ensure an adequate water supply. The city provided a system of basins for industrial users at the Fonte Follonica, the Fonte Peschaia, the Fonte Ovile, and the Fonte Nuova. As previously noted, none had as much water as the Fonte Branda, which in turn is why that fountain appears most frequently in the statutes.[27]

The process of making woolen cloth in the Middle Ages included numerous steps, beginning with the shearing of sheep and the carding of wool, which took place in the countryside; to spinning yarn, which was put out to individual households; to the dying and fulling of woven cloth, which depended upon centralized facilities. The last two activities were more closely regulated because they depended upon the shared use of publicly owned facilities. The guilds often managed and maintained the large basins by the fountains and thus had an interest in regulating their members' use of the facilities. It appears that the guilds in Siena also owned, or at least completely controlled, some of the facilities. However, Sienese guilds managed and maintained basins near the fountains by the same means as the city: the promulgation of written statutes, inspections, secret informers, and fines. Examples of the statutes include the *Statuto dell'Arte di Carnajuoli di Siena* (Law of the Butchers' Guild), the *Statuto dell'Arte di Lana di Siena* (Law of the Woolworkers' Guild), and the Law of the *Arte di Cuoiai e Calzolai* (Guild of the Leatherworkers and Shoemakers).[28]

The Law of the Woolworkers' Guild, dating from 1298–1309, provides an excellent outline of some of the processes the guild sought to regulate. For instance, the rubric entitled "In what manner one should wash in the pool of the Woolworkers' Guild" tells what types of cloth and what colors can be combined in each of three pools below the Fonte Branda:

> Item, we constitute and order, that no member of the Woolworkers' Guild of Siena may nor ought wash in the upper pool wool nor fine carded wool nor cloth, if not wool or carded wool, nor vermillion and green cloth and yellow and blood red and dyed with orchil, orange, and wool, and fine carded wool treated with alum, and cotton wool that has been treated with alum in these colors. One may also wash in said pool cloth that has been dyed in the wool or raw cloth, and all woolen cloth of every color; except cloth dyed in azure vegetable dye or black, and woolen cloth and cotton wool dyed in indigo, and all cloths and cotton wool that has been [treated in the manner referred to as] "infolliated." In the second pool one may wash the wool and fine-carded wool dyed in woad, and cloths that are dyed in woad, and all colors of woad, and no other colors. In the third pool one may wash unwashed wool blends, and black ones, and black cloths, and every shade of black or indigo, and infolliated cotton wool, and low quality black cloths. In the new pool down below, one may wash raw unwashed wool and one may soak skins and leather, but not in any other; and [perform] all the tasks related to tanning.[29]

Obviously, a great deal of thought and practical knowledge went into working out the specifics of which operations could be performed upstream of others. The mention of a "new" pool suggests that the arrangements evolved to reflect changes in the way crafts were practiced. The statute continues with details of the fines to be paid for any violation of the rules.

Readers will recall the two descriptions of the Fonte Branda complex from the previous chapter. The first published around 1556 and the second in 1625, supplement the regulatory prescriptions reproduced above by providing narrative portraits of the many functions the Branda complex performed. Though the descriptions date from as much as two centuries after the period covered by this study, they remain useful because they correspond well to the prescriptions of the Woolworkers' Guild. The exact location of these mills will have to await archaeological research, but if we can use these accounts to make inferences about physical arrangements two centuries earlier, it would appear the legal prescriptions and layout of the fountains worked together to make the most of a limited water supply.

It is possible that some of the mills mentioned above were used for fulling. It would make sense to concentrate textile operations geographically to reduce transport costs. After being woven and dyed, woolen cloth was fulled to shrink the cloth and thus interlock the burrs on the strands of wool to make a tight, almost waterproof product. E. M. Carus-Wilson summed up the requirements of the fuller's trade as follows:

> Fulling was a process that demanded not only skill and a certain amount of equipment but also a considerable space of open ground for drying and a plentiful supply of clear, fresh water. From this it followed that fulleries were often set up along a watercourse in close proximity to each other and that there was need for regulating the use of the water. This may well have been one of the reasons for the frequent formation of guilds among the fullers, in Roman as in medieval times.[30]

The Sienese statutes do not reveal the extent to which fulling was carried out manually inside the city compared to the proportion of woolen cloth transported into the countryside to be fulled at water-powered mills. However, the number of statutes governing the transport of cloth to the fulling mills and regulating the behavior of the fullers suggests that much of the cloth was fulled at mills outside the city. For instance, one rubric ordered that the officials of the guild urge the city to finish a road it had begun to build, because completing the road to Valle di Querceto would be "useful for going to our fulling mills." The succeeding rubric ordered the officials of the guild "to send two men secretly in the six-month term of their office, that is, one time in the first three months, and the other in the following three months, to each fullery to make sure that the guild's things are maintained in a condition according to the written agreements." All nineteen rubrics of the seventh chapter of the Laws of the Woolworkers' Guild are devoted to the use and operation of the fulleries, suggesting that the application of water power was the norm.[31]

One can see in the city's statutes an attempt to accommodate the various, often-competing requirements of the wool industry. One statute specified that the overflow from the aqueduct feeding the Fonte Branda be used to give the wool industry a larger supply of water. Another required that wool workers use only the basins allotted to them. They could not wash wool in the laundry basins, horse troughs, or the pool called a *guazzatoio* (apparently designed for watering or immersing livestock, especially horses) of the Fonte Branda. A provision dating from 1298 allowed dyers to dig their own little drains into which they could dump the water polluted with dyes and mordants. The channels could be a brick wide at most, pass under the streets, and be covered by brick. Through numerous such statutes, reiterated and sometimes refined with each new edition of statutes, the city and its guilds attempted to create a system of resource allocation that would maximize the utility of scarce collective resources and protect the urban environment by closely legislating the behavior of individuals and corporations.[32]

The waste produced by slaughterhouses, butchers, and tanners posed an ongoing menace to the fountains, streets, and residents of the neighborhoods in which they operated. Many of these messy operations took place around the Fonte Vetrice (no longer extant). Although its exact location has not been determined, the Fonte Vetrice was most likely situated on the south side of the Val di Piatta, across from the Fonte Branda, a site now occupied by abandoned nineteenth-century slaughterhouses. It would appear that textile-related industrial processes clustered around the Fonte Branda and those activities concerned with meat processing centered around the Fonte Vetrice.[33]

William Bowsky used the ongoing struggle between the butchers and the city to illustrate the ways in which the government attempted to regulate very powerful interests and to find a compromise that would satisfy both members of the trade and the public good. He documents a see-sawing pattern of implementing regulations and securing their repeal. He found that although the butchers themselves had some political standing through their guild, they exercised a degree of influence out of proportion to their place in the city's economy and body politic. Bowsky concluded that their influence stemmed from their powerful allies, the large landowners, who raised the livestock for sale as meat. That sector of the agricultural economy remained in the hands of some of the oldest, most powerful families in the region whose wealth still largely depended on their vast farms, encompassing the rich pasture land south and west of the city. Perhaps the perception that their fortunes were in decline, relative to new ways of making money, made the rural magnates particularly resistant to the yoke of urban government.[34]

Siena's statutes suggest that residents found butchering to be one of the most offensive activities associated with the use of water. Regulations prohibited a range of activities including throwing intestines in the street in front of the Fonte Vetrice, from the city gate up to the fountain, anywhere within about fifty meters of other fountains; or throwing intestines or putrid things; or skinning or butchering any animal or carcass in those places. Another law forbade anyone to throw, or make anyone else throw, any filthy or putrid thing or carcasses or blood, by day or by night, from the fountain of the Val di Montone and from above, by any means. The same law prescribed the means by which it was to be enforced, ordering that the Fonte di Val di Montone be guarded at the expense of the city. Posting guards, though not universal, was a fairly typical means of discouraging the abuse of the fountains. Entries in the *Biccherne* (the city account books), show frequent payments for the services of various guards, proving that at least some of the regulations were in fact put into practice.[35] For instance, an entry for June of 1310 reads:

Also, £3 to Biagio Brandi, guard of the Fonte Branda.
Also, 10s. to Chambio, guard of the Fonte Follonicha.
Also, 10s. to Gíovanino, guard of the Fonte Uvile [Ovile].
Also, 10s. to Gualfreduccio, guard of the Fonte Pescharia.
Also, 10s. to Bernarduccio Palmieri, guard of the Fonte Val di Montone.
Also, 10s. to Lolo Mini, guard of the Fonte Vallerozi.
Also, 10s. to Mino Muschio, guard of the Fonte of the Romiti Friars.[36]

The huge difference in the fee paid the guard of the Fonte Branda suggests that the scope of his job was much wider than that of the other guards, which is consistent with what we know about that fountain's extensive industrial facilities. The decision to post a guard may also reflect the economic value of those facilities and the potential cost of any disruption to their operation.

Regulations also governed activities related to butchering, such as the production of leather. Turning animal hides into leather entailed several steps. Each step had the potential to spread filth in the streets, pollute the water, foul the air, attract vermin, and create a general nuisance. Statutes pertaining to the first stage in the process, skinning the carcass, have already been discussed. After flaying the animal it was necessary to remove the hair from the outside of the skin and any remaining flesh from the inside. Victoria Gabbitas explains the process of removing hair and flesh "by soaking the hide in a lime solution and then scraping both sides with a knife. The lime also helped to separate the fibers of the hide, which allowed better penetration of the tanning solution." Another rubric attempted to regulate the mess caused by scraping the hair from the skin, ordering that no one throw any slaked lime with or without any hide, or any flakes of lime, or hides into the large *guazzatoio* of the Fonte Branda. Tanners were forbidden to put skins in the laundry basins, horse troughs, or *guazzatoio* of the Fonte Branda. The statutes also prohibited activities that were hidden from the public eye when they threatened a public resource. For example, tanners who worked in the cellars, the ones above the basin and trough of the Fonte Branda, were not permitted to skin or soak or tan or to make someone else put any skin or leather in the immediate vicinity of the Fonte Branda. The city authorities seem to have understood that by-products from leather production constituted at least an aesthetic or economic hazard whenever there was a chance of their entering a fountain. Regulations attempted to ensure that the filthiest operations were performed downstream of all the others so that the effluent would run into a ditch away from other fountains and basins, eventually passing beneath the city walls and soaking into the ground or flowing into streams below the city.[37]

From water as a means of transporting industrial waste we turn to water as a means of extinguishing fires. Because of their frequency and destructive potential, city officials and lawmakers gave considerable thought to the prevention of fire and firefighting. Cities attempted to prevent fire by regulating the use of flammable building materials such as wood and straw and by locating crafts that used fire at the perimeters or outside city walls. The buildings of fourteenth-century Florence were more expensive, more permanent, and comprised of more fire-resistant materials than those which they replaced and the same was true in Siena. Firefighting began with transporting water from fountains to burning buildings by means of bucket brigades composed of civilian volunteers and paid porters and teamsters. A statute of 1282 set the rates at which water carriers would be reimbursed by the city for each bucket carried at 3d per tub of water. The provision of very large basins in each of the fountain complexes allowed many people to simultaneously dip a bucket or a jar in the water rather than having to wait in line for the trickle from a single spigot. If a fire grew out of control, which it often did, members of the building trades would tear down nearby buildings to isolate a burning building from the rest of the city. For example, a fire raged for two days in 1302 and efforts to extinguish it involved 241 masters in the building trades. A fire in 1307 destroyed twenty-nine houses and the city reimbursed 164 people for their work putting it out. Through a combination of preventive measures such as zoning and the substitution of building materials and the encouragement of remedies such as the organization of volunteer fire fighters, the city was able to gradually reduce the risk of fires in the city. These limited means constrained the effectiveness of urban fire control used until the advent of portable mechanical pumps in the seventeenth century.[38]

The personal use of water by individuals as a beverage or for cooking, laundry, and bathing proves to be the most elusive category of urban water use. With the exception of some statutes ordering that fountains or wells be built for the convenience of travelers, the statutory silence on this subject might tempt one to conclude that people in the Middle Ages never drank water. Prescriptive literature from the period suggests otherwise. For example, Alberti wrote:

> Since a city requires a large amount of water not only for drinking, but also for washing, for gardens, tanners and fullers, and drains, and—this is very important—in case of sudden outbreak of fire, the best should be reserved for drinking, and the remainder distributed according to need.[39]

Alberti's advice strongly suggests that people not only drank water, but that they thought about ways of guarding its quality and preserving the best for human consumption. From the amount of attention he pays to obtaining potable water, it is clear that he considered it an important beverage. Alberti draws upon the writings of the ancients, going into considerable detail to describe the best types of drinking water. He cautions his readers against resorting to an inferior water supply by describing its potentially debilitating health consequences. The consumption of water as a beverage was implicitly among the reasons for bringing water to Siena, and explicitly at the head of the hierarchy of uses, as regulations ensure that drinking water was the single best protected type of water. The silence of the statutes likely indicates only that the consumption of drinking water led to few abuses for which there was a legal remedy.[40]

Although the city provided specific public facilities in which to do laundry, the process of washing clothes is only slightly more visible in the statutes than the activity of drinking water. The statutes often made reference to the fountains' *lavatoi* (laundry basins). Most often the laws protected the laundry basins from being abused by woolworkers or leatherworkers. For example, one law ordered that neither wool nor leather be put into the Fonte Branda's laundry basins. Another statute ordered the construction of a laundry basin of adequate width so that women could comfortably wash at the basin. The statutes did not attempt to regulate how laundry was done; they only provided access to facilities in which to do it. Does this indicate that the laundresses were self-regulating to a greater extent and better behaved than the woolworkers? Or was doing laundry inherently less likely to contaminate the rest of the water supply? The physical arrangement, with the placement of the laundry basins right beneath the needs of animals, suggests that the effluent from washing clothes was seen as harmful to humans and livestock but not to Siena's industrial users of water.[41]

The history of laundering clothing, from the Middle Ages until the Industrial Revolution, is virtually an untouched subject. A search of the secondary literature has revealed very few works on the subject of laundry in the premodern world. In fact, modern notions of laundry, defined by actual washing of garments, may not have applied to most clothes in the Middle Ages. It remains to be discovered what procedures were actually used. Most research has concerned the mechanization or industrialization of laundry in later centuries. Apparently the manual labor of women, especially drudgery, has interested few historians. If surviving records are any indication, such tasks did not interest contemporary observers either. Changes in the methods, facilities, soaps, use of lye, and drying methods can only be surmised.

Figure 4.3 Detail of *lavatoio*, Fonte Ovile around 1900. From Bargagli-Petrucci, *Fonti di Siena.*

Regional variations and the rate of modernization are also completely unknown. One cannot overstate the difficulty in making inferences from such slender, almost non-existent information. Although early twentieth-century photographs show now-destroyed laundry basins still in use, it is impossible to tell from the images when the basins were built or last modified. Other photographs show linens spread out on grassy hillsides near the city, where they could both dry and be bleached by the sun. Photographs such as these offer tantalizing clues as to what could have been common practice earlier. One cannot rule out the possibility that the laundry basins were originally built for some other purpose and then abandoned to the laundresses later. Only further research will show how safe it is to draw any conclusions about medieval and early modern laundry practices from such late sources. The Sienese were not the first in Italy to distribute water according to a hierarchy of quality to conserve scarce resources. Frontinus' description of each of the eleven waters brought to Rome by the aqueducts shows a similar system of allocation of hydrologic resources according to quality, a point discussed in Chapter 3.[42]

Towards the end of the twentieth century, at what was once the eastern edge of the Roman Empire, what is now Yazd province in Central Iran, archaeologist Susan Roaf observes, "the same water may be used for up to five different functions in the course of its journey through the village." She adds,

Figure 4.4 Women doing laundry in the Fonte Nuova *lavatoio* around 1900. It is likely that the laundry basin used to be outside the fountain and below it. From Bargagli-Petrucci, *Fonti di Siena.*

Figure 4.5 Interior of *lavatoio* complex connected to Fonte Branda, around 1900 (no longer extant). This was the city's largest and most elaborate facility. From Bargagli-Petrucci, *Fonti di Siena.*

"this practice of water conservation and re-use is associated with a strict hierarchy of uses along the main street/water course, a hierarchy determined by the varying quality of water required." Other Mediterranean and Roman examples of the hierarchical use of water follow the same principle. Like other cities, Siena embraced such a hierarchical arrangement for the use of its multiple water supplies. In Siena, the parallels to the patterns in Yazd Province and the prescriptions of Frontinus are most conspicuous along the watercourse and road below the Fonte Branda, but they seem to apply to all the fountain complexes.[43]

Similarities between the Sienese water supply system and those of the ancient Near East should come as no surprise if for no other reason than the seasonal scarcity of water that pervades both regions. Using multiple sources of water conferred several advantages upon the citizens of Siena, foremost among them being increased reliability through redundancy. Dora Crouch, in her study of ancient Greek water management, argues that redundancy of water supply offers increased security in the face of the failure of any one subsystem and that such redundancy is a feature of Greek cities in the Mediterranean. The use of overlapping water supplies is likely to be common to all urban civilizations coping with the Mediterranean basin's pattern of unevenly distributed annual rainfall. The perennial threat of siege warfare, during which an enemy might try to sever a city's supply of food and water, also demanded a backup system. Crouch adds that a concomitant aspect of the Greeks' strategy included the categorization and use of water according to its quality, reserving the scarcest commodity, potable water, for drinking alone. The Sienese employed a similar system of allocating water to users according to quality. Siena's strategies for averting water shortages in the face of uneven supplies show themselves to be typical of Mediterranean cities.[44]

In conclusion, there are four major points to consider: the extent and success of regulation; the nature of sanitary conditions; the role of water in the industrialization of craft processes; and the hierarchical nature of the system of water supply. The rulers of medieval Siena put considerable effort into regulating the use of its scarcest resource. However, the extent to which regulations were obeyed or enforced remains unknown. Relevant records of criminal proceedings, if any survive, have yet to be found. The extensive delegation of powers and the voice given commercial interests in the formulation of city and guild regulations suggest that many powerful factions had an interest in cleanliness and order. The delegation of powers to committees of citizens had reciprocal effects. It gave neighborhoods some say in the decision-making process, and it extended the reach of the city into those neighborhoods through paid guards and informers. Presumably such widespread participation helped build consensus, resulting in laws that were more easily enforced than if they had been imposed from above.[45]

Lynn Thorndike's and Lewis Mumford's estimations of medieval cleanliness, made earlier this century, appear to be borne out by Sienese evidence. If not the standards of practice, then at least the prescriptions for Sienese behavior "were far more adequate than most Victorian commentators—and those who still echo their prejudices and blandly repeat their errors—believed." Without comparative evidence from other cities, it is difficult to discern precisely how typical or exceptional Siena was in the Middle Ages. The few clues we have, such as the statements about the importance of urban sanitation from Florentines Alberti and Bruno, suggest that cleanliness was much on the minds of at least Siena's most powerful neighbor. It is likely that Siena may turn out to be exceptional with respect to its surviving documentation, artifacts, and entrancing beauty, but that its sanitary practices and regulations were typical of at least Tuscany, as well as some other Mediterranean cities, and perhaps the starting point for a trend that was spreading throughout Europe.[46]

If such is the case, then it becomes easier to situate Alberti's scathing remarks about the stench of open sewers in Siena. He said, "Not only does the whole town stink, at the beginning and end of the night watch, when refuse receptacles are emptied out of the windows, but during the day as well, it is filthy and offensively vaporous." Coming after all the rhetoric of beauty and cleanliness, his remarks seem incongruous at first. His observations could have been based on what he observed during his visit to Siena in the service of Eugene IV. He remained in the city for seven months in 1443, well after the city's demographic expansion had filled in much of the green space with buildings. Even if his words are intended to display his loyalty to Florence, his choosing cleanliness as a point of comparison between the two cities says much about attitudes toward filth in his day. Although Siena's population had fallen precipitously after the Black Death in 1348, the denser physical structure of the city lay largely intact. It is perhaps possible that Alberti saw and commented upon a greatly compromised waste disposal system—one that was not being maintained as those who built and regulated it had intended.[47]

Water played a crucial role in urban crafts, but many of these very crafts also posed a risk to the water supply. Through municipal statutes and self-regulation by the guilds, Siena attempted to promote the interests of the guilds and to reduce the negative impacts of craft activities. The limited supply of water made careful allocation all the more important, not only to extend the supply, but also to reduce potential conflict among users. Ultimately only the most essential or highest value-added tasks, such as laundering, wool washing, and butchering, could be performed within the city walls. All other industrial uses were consigned to rivers well outside the city, despite the increased costs of transport and operation.

Finally, the principle of Sienese water regulation was a hierarchical division of water uses, both in law and in the physical form of the structures by which water was distributed. Dividing water use by quality and type goes back to ancient Greece and Rome and is likely to have been practiced wherever there was at least seasonal scarcity. Such a hierarchal division of water provides a powerful and useful model for allocating contemporary water resources, instead of using water once and then disposing of it entirely. The hierarchical structure in Siena is not unique. The reuse of water echoes descriptions of systems in Cistercian monasteries. And the Sienese system mirrors traditional ones as old as Siena's, some of which are still in use in more arid regions of the world, such as the province of Yazd. Lessons learned from the history of water supply systems at Siena, in other parts of the Mediterranean, and in the Middle East can provide a model for systems in the developed and developing worlds today. By using and re-using water, the total volume of water required per capita is reduced. Formulating standards of water quality appropriate to the task at hand can allow cities to avoid squandering potable water on flushing toilets and washing streets.[48]

Chapter Five

Antecedents, Parallels, and Diffusion

The medieval aqueducts that have supplied Siena with water for nearly a millennium depended upon earlier systems and may have also borrowed from contemporaneous developments in other regions. In turn, the process of constructing and maintaining Siena's water supply system resulted in the accumulation of a large body of technical know-how. This accumulated experience provided a body of knowledge that informed future generations of engineers. The Sienese system has historical antecedents dating back to Roman times and very likely to the Etruscan civilization that Rome conquered as it expanded north into Etruria. Among these antecedents are the *cuniculi* (underground drainage tunnels carved through rock) of the Etruscans, at places such as Veii. They also include the subterranean aqueducts of the Romans, supplying Rome and other cities such as Florence and Bologna. In addition to technologies that preceded it, the Sienese system resembled contemporaneous systems that worked on analogous principles. Parallels include the adits used to drain mines and supply water in Bohemia and Bavaria and the tunnels by which military sappers destroyed fortifications. The most striking similarity, however, is found in the qanats that watered fields and cities across the shores of the Mediterranean, in the Levant, North Africa, Sicily, and Spain.[1]

The question as to how or when the technique of qanats reached Siena is not yet answerable. There are four major possibilities: independent invention in Siena and at least three waves of diffusion. Thomas F. Glick uses the phrase "shared technologies" to describe those agrarian and industrial techniques which were diffused through the Mediterranean world in antiquity. He includes both pre-Roman and Roman techniques in this description, accounting for the first two waves of diffusion. The transmission of ideas and techniques from China and India via the Islamic world during the Middle

Ages, of which Joseph Needham, Lynn White, Jr., and Glick have written, accounts for a third wave of diffusion. Insofar as this technique of qanat building reached the shores of Italy at least three different times since their appearance in Iran during the first millennium B.C.E., it becomes difficult to make the case for independent invention. Thus, we should turn our attention to models of diffusion.[2]

MODELS OF DIFFUSION

In the absence of documents attesting to the transfer of technology, the historian must depend upon physical evidence of the diffusion of a technology, using methods addressed early in the twentieth century in the anthropological literature. A. L. Kroeber defined diffusion in 1923 simply as "borrowing"—one people learning an institution or belief from another. Kroeber later coined the phrase "stimulus diffusion" to describe a sort of incomplete transfer of a cultural fragment such as an idea, technique, or technology from one culture to another. He argued that in the hands of skilled observers such as artists, builders, and engineers seeking models on which to base their designs, technological artifacts can be the vehicle of transmission of technical knowledge, despite discontinuities in the social and political structures under which they were built. Indirect communication, such as that afforded by rumor, could be a stimulus to innovation, as well. Although much of Kroeber's work has been ignored by other anthropologists, historians and sociologists have readily adopted the idea of the stimulus diffusion model to explain the transfer of technology across groups of users.[3]

The concept of stimulus diffusion has proven particularly useful for investigating those technologies and periods that have left few written records, such as the Middle Ages. Toward the end of his career Lynn White explicitly acknowledged his debt to Kroeber for introducing him to the methods of anthropology when he wrote about his first encounter, in 1933, with Kroeber's 1923 textbook, *Anthropology.* He noted that "Kroeber searched for depictions of things, for surviving artifacts, and for the total debris of the past." The "debris" formed the very sorts of evidence in which Kroeber was to later assert that traces of stimulus diffusion could be found. As a result of his encounter with Kroeber's work, White continues, "I determined to see what I might do to apply the methods of Cultural Anthropology to the Middle Ages." By adding the analysis of artifacts to the use of documents, White made one of his most significant contributions to the study of medieval technology.[4]

Other historians have acknowledged the utility of some of the methods of cultural anthropology by adopting its methods and models. Joseph

Needham embraced the stimulus diffusion model when he addressed the issue of transmission from China to the West, suggesting that:

> A simple hint, a faint suggestion of an idea, might be sufficient to set off a train of development which would lead to roughly similar phenomena in later ages, apparently wholly independent in origin. . . . Or the news that some technical process had successfully been accomplished in some far-away part of the world might encourage certain people to solve the problem anew entirely in their own way.[5]

Needham and White helped to make the use of the diffusion model routine in the study of medieval technology. The idea of diffusion of technology has become so widely accepted that few historians feel the need to defend its application.[6]

The idea of diffusion and its corollary, stimulus diffusion, have gained widespread acceptance in the history of archaeology, technology, the sociology of technology, and allied disciplines such as economic botany and ethnobotany. In fact, the uniqueness of biological species has likely made the acceptance of the diffusion model that much easier. In 1975, for example, White argued that "water buffaloes couldn't be invented twice." The increasing amount of evidence gathered since the 1930s has further eased the burden placed upon the diffusion model as pieces along the route of transmission gradually accrue and are incorporated into more recent scholarship. Nevertheless, the recognition of Islamic influences on European technology has been hindered by various cultural prejudices. Most students of European history still insist on an uncritical acceptance of the "Renaissance" model in which early modern Europe is simply Rome reborn. Whereas, historians of science have long accepted the diffusion of ideas from East to West, this way of thinking has been slow to catch on in other specialties.[7]

DIFFUSION FROM EAST TO WEST

The entire discussion of Arabic influences on western culture has been clouded by prejudices stemming from the time when the Christian West lagged behind the Islamic world culturally, scientifically, and economically. Prejudices about the nature of the later Middle Ages are traceable to the Italian Humanists such as Petrarch, who wrote in 1370 to his friend the physician and clock maker, Giovanni de' Dondi, "I will not be persuaded that anything good can come from Arabia. . . ." As White notes, this anti-Arab attitude was compounded by the European chauvinism typified in Francis Bacon's *Novum organum* (1620) in which he presents major components of the European inheritance from the Islamic world as independent discoveries typifying European genius. These discoveries include cannon, silk, the nautical compass, sugar, paper, and printing.[8]

Historians of science, beginning with Charles Homer Haskins and George Sarton, have long appreciated the role of Arabic learning and scholarship in preserving the texts of Aristotle and other ancient authors, as well as making many original contributions. By A.D. 1000 almost the entire corpus of Greek medicine, natural philosophy, and mathematical science had been rendered into usable Arabic versions. Historians have also recognized that the commentaries of the Islamic scholars Ibn Sina (980–1037) and Ibn Rushd (1126–1198), known to Christians as Avicenna and Averroës, "were as important as the works of Aristotle himself in forming European scientific and philosophical thought." In the field of mathematics Muslims introduced not only "the Indian system of reckoning using the nine Arabic numerals with a point for zero" in the ninth century, but also the use of decimal fractions in the tenth century. In the field of astronomy Muslim scholars preserved the Greek works of the school of Alexandria, including Ptolemy's *Almagest.* They also translated Indian works such as the *Siddhanta* beginning in the eighth century. Their own achievements "lay in the fields of observation, instrumentation, and the development of spherical trigonometry for the solutions of problems in astronomical mathematics." Their best known work in physics is on optics and the nature of light. The *Book on Optics* by Ibn al-Haytham (sometimes Latinized as Alhazen, 965–1039) "established that rays of light start from the object and travel towards the eye, not the reverse as the Greeks had believed." This is a small representative sample of Muslim contributions to learning that also include medicine, chemistry, architecture, and other fields.[9]

Despite these contributions, the predisposition to believe in the "European-ness" of the Renaissance persists. This mistaken perception is easy to support because most sources that would disprove it remain uncatalogued, unpublished, and untranslated. Most western medievalists' lack of Arabic has only served to increase this perception. The task of studying medieval technology is sufficiently daunting already without asking each student to add Arabic to her toolkit, although this might be an ideal solution for the next generation of scholars. Many of the catalogues that are available, like Michele Amari's *Biblioteca arabo-sicula,* require Arabic. Even using the English-language *Encyclopedia of Islam* requires one to learn the transliteration of the Arabic terms under which each entry is alphabetized. Such linguistic difficulties make it easier to ignore or to "forget" Islamic and Far Eastern influences.[10]

More than forty years ago, Charles Singer concluded the second volume of his *History of Technology* with an epilogue entitled "East and West in Retrospect." He wrote:

> Europe, however, is but a small peninsula extending from the great land masses of Afrasia. This is indeed its geographical status and this, until at least the thirteenth century A.D., was generally also its technological status. In skill and inventiveness, during most of the period treated in this volume, the Near East was superior to the west, and the Far East perhaps superior to both.[11]

Singer argued that too much emphasis has been placed on the role of Greece and Rome; he highlighted the roles of major agents of diffusion, such as pilgrims, crusaders, and traders; he indicated the importance of technological and scientific advances arriving from India and China; and he noted the locations and significance of trade routes within Europe and between Europe and the Far East.

In his review of the volume, White predicted that Singer's epilogue would be "the most widely read chapter" of the monumental series. But White cautioned, "This may be unfortunate; for his chief theses are highly debatable in terms of the materials presented in the rest of the volume." White argued for greater appreciation for the technological inventiveness of Greece and Rome, citing the Antikythera machine, about which Derek J. de Solla Price had recently published, as well as noting classical contributions to artillery, glass blowing, and the Vitruvian mill. Nor was White ready at that time to accord the medieval civilizations outside of Europe the credit for so many fundamental contributions. Singer, on the other hand, relying mostly on Needham, was willing to recognize contributions from abroad. Ironically, in the next twenty years, White would do as much as any scholar to confirm Singer's thesis about extra-European influences.[12]

Unfortunately, the specialized work needed to confirm or disprove Singer's synthetic look back has barely gotten underway. Instead of the reception White predicted, Singer's essay has been largely ignored. Almost thirty years later, two of the scholars who have done the most work in the directions Singer indicated, Ahmad al-Hassan and Donald R. Hill, while bringing many details to light, ultimately could do little more than quote Singer's conclusions and add a few examples in support of Singer while decrying the lack of evidence then available to scholars. Other scholars focusing on relations between Christian and Islamic civilizations in the Mediterranean face similar frustrations. John H. Pryor, in his study of maritime history, bemoans the lack of Byzantine and Muslim sources. Abraham Udovitch regrets the lack of primary commercial documents, noting that almost none has survived. The situation for medieval Iberia has recently improved with Olivia Remie Constable's publication of sources in a recent collection. Although Singer decided that economic history fell outside the scope of his series, these studies demonstrate that he was correct in concluding that "Trade-routes—like exchanges, commerce, and book-keeping[—] . . . are intimately linked with our subject." Some of the best examples of these links come from S. D. Goitein's pioneering work, in which he used documents from the Cairo Geniza that other scholars had long ignored, to write an economic history. The Geniza contained thousands of medieval documents that testify to numerous connections across the Mediterranean. Thomas F. Glick, in his study of medieval Spain, looks at technology such as the noria, a wheel with buckets for automatically raising water; deep wells; pottery; and glass. His research shows that the flow of ideas and techniques was from East to West.[13]

Joseph L. Strayer made one of the strongest commitments to Singer's view when he attempted an unprecedented integration of scholarship regarding the Christian and Islamic worlds of the Middle Ages in the *Dictionary of the Middle Ages.* Still, most articles in the *Dictionary* reflect the divide in scholarship by addressing topics from either a "western" or an "Islamic" point of view; a notable example is Lynn White's entry, "Technology, Western."[14]

White described the obstacles confronting students of western European medieval history in his 1972 essay, "The Flavor of Early-Renaissance Technology," and it is worth reviewing what he said then in light of the need to attain a more comprehensive view of the Mediterranean context in which western European technology flourished in the later Middle Ages. White noted the "dearth of scholarly work in the field," identified the chief need to be "the careful publication of the notebooks of the engineers of the period," and called for the publication of editions of manuscripts by Taccola, Guido da Vigevano, Filarete, Francesco di Giorgio, Giovanni Fontana, and the Anonymous of the Hussite Wars. Except for the work of Taccola, an Italian edition of Francesco's Trattati, one of Fontana's ten known manuscripts, and Bert Hall's wonderful edition of the Anonymous of the Hussite Wars, we still lack some of the most fundamental building blocks of any inquiry of western medieval technology. Finally, White lamented that "few scholars combine the necessary paleographic and linguistic skills with technical insight."[15]

The lack of scholarly editions is especially unfortunate in the case of Fontana (ca. 1390-after 1450) because the few diagrams that have been published from his notebooks reveal tantalizing stylistic similarities between Fontana's drawings and those of the Islamic engineers of the tenth through thirteenth centuries such as the Banu Musa (fl. ninth century), al-Jazari (twelfth century), al-Ramah (d. 1294), and al-Dimashqi (d. 1327). In his 1971 *Physis* article, Frank Prager noted that "some of Fontana's drawings reflect Arabic source material of uncertain authorship" but that the lack of published Arabic sources made it impossible to continue the identification further. The possibility of influences is made all the more plausible when we see the number of Arabic authors known to be among Fontana's sources. More recent scholarship by Eugenio Battisti and Giuseppa Saccaro Battisti reveals that in addition to al-Kindi (mid-ninth century), al-Biruni (d. after 1050), and certain "unspecified Greeks, Persians, and Saracens" noted by Prager, Fontana directly cites more than twenty other Arabic and Hebrew sources. Fontana's access to so many extra-European sources is significant because he was a contemporary of Taccola, working in Venice. Whereas the lack of editions for fundamental European manuscripts is bad enough, Hassan and Hill inform us that the situation for Arabic manuscripts is even worse: "only a handful" of manuscripts have been published, whereas thousands of others have not even been properly catalogued.[16]

Nonetheless, there is a long history of trying to dismantle these preconceptions. Since the 1920s, scholars such as Charles Homer Haskins and Wilhelm von Heyd have tried to encompass the far shores of the Mediterranean in their respective histories of science and commerce in the late Middle Ages. Heyd gives us a sense of what he called "an extraordinary movement of travelers between Europe and Palestine," a list that included, in addition to the expected crusaders, ecclesiastics and knights, pilgrims and merchants, adventurers and manufacturers. In his book of the same name, Haskins proposed a "Renaissance of the Twelfth Century," a "Renaissance" as much in debt to newly imported ideas as those revived from antiquity. He also recognized the role of trade and commerce in transmitting ideas but did not fully develop the theme.[17]

Economic and business historians have worked since the 1930s to show the transnational character of trade routes, goods, merchants, and the interplay between markets in Europe and beyond. In his edition of Francesco Balducci Pegolotti's *La Practica della Mercatura,* Allan Evans brought the modern reader the first complete, authoritative edition of this important early fourteenth-century handbook for merchants. In addition to presenting Pegolotti's advice on purchasing hundreds of commodities, Evans collated the material in an annotated list of trade goods. Florence Edler gave us another equally important source in her *Glossary of Medieval Terms of Business.* In 1956, Irving Raymond and Robert S. Lopez filled a huge void in materials available to English-speaking students of medieval trade with their publication of a careful, wide-ranging selection of documents and commentaries in their *Medieval Trade in the Mediterranean World* that remains unsurpassed.[18]

The task of synthesizing the findings of historians of commerce and science has only begun. In his "Epilogue," Singer reproduced a time line—based on Joseph Needham's work—that showed the date of first appearance of numerous devices in the East and West. Forty years later, this chart remains a useful guide for historians of medieval technology. Since then, through the ongoing publication of *Science and Civilization,* Needham and his team of researchers have provided a steady supply of new data to historians. White acknowledged Needham's contribution, praising him as being "learned in medieval European technology to the point where no medievalist can afford to ignore him."[19]

S. D. Goitein's publication of economic and commercial documents from the Cairo Geniza, beginning in the 1950s, provided immensely rich sources, as well as a new perspective to historians, as thousands of documents began to reveal the details of day-to-day life and commerce among Christians, Jews, and Muslims in Fustat and the surrounding world. Goitein's research supports Singer's position on the relative backwardness of Europe. Goitein finds that "to sell to the Europeans was the dream of the local merchants," that "Europeans were expected to be satisfied with second-rate merchandise," and that as late as the year 1060 Europeans were still unable to distinguish quality goods or to know prices. Abraham Udovitch showed that what was thought to be one of the seminal inventions of Italian merchants, the *commenda* contract, was more likely to be a borrowing from their Muslim trading partners. W. Montgomery Watt has synthesized much secondary material, making it available to an English-reading public for the first time in *The Influence of Islam on Medieval Europe.* Perhaps Watt's most valuable contribution is his eye-opening glossary of Arabic loan words, a list that not even the second edition of the *Oxford English Dictionary* has fully integrated into its etymologies.[20]

The work of Heyd and Haskins began in the 1920s and continues to expand and to touch other disciplines. The 1980s saw the fertile collaboration of Ahmad Y. al-Hassan and Donald R. Hill in their UNESCO-sponsored book, *Islamic Technology: An Illustrated History.* María Rosa Menocal, a historian of medieval literature, extended and developed the debate that Américo Castro and others had begun about the "European-ness" of those writers who most fully embodied the idea of the Italian Renaissance, namely Dante, Petrarch, and Boccaccio.[21] Not surprisingly, she found the same pattern in literary borrowings from Arabic literature that economic historians had found with regard to trade goods. A reader comparing the practices of

Italian merchants and writers might speculate about the nature of "import substitution" extending to the literary marketplace as Europeans learned to be troubadours, much as they had learned to be glassblowers, silk weavers, and paper makers.

As a result of the disciplinary gulf among scholars of the Christian, Jewish, and Muslim Middle Ages in the Mediterranean world, the process of answering many fundamental questions about the nuts and bolts of medieval technology in Europe is going very slowly. Except for White's contributions to our knowledge in the 1960s and 1970s, the field remains stalled where it was in the 1950s. For instance, despite the publication of three monographs on wind power in the last ten years, none has added to what has long been known about the first appearance of vertical-axle windmills in the West.[22]

Although scholars have long debated whether the cannon, silk, nautical (magnetic) compass, sugar, paper, printing, troubadour poetry, and elements of mechanical clockwork have their origins somewhere outside Christian Europe, we are left with inferences, hypotheses, and polemics about why this or that item must have been independently invented or could not have been diffused. The gulf among historical subdisciplines becomes clear when one realizes that few scholars other than economic historians and historians of science and technology have found it significant that the first known instances of many of these items are often at points of contact between the two cultures, such as European and foreign ports, frontiers in Spain, newly conquered lands in the Levant and Spain, or the multilingual courts of the Hohenstaufen in southern Italy and Sicily. I think the factor of diffusion has so long been ignored because it has been invisible to those historians whose studies of daily life in places seemingly remote from the Islamic world, like England, fail to reflect how strong the links were in the medieval economy formed between London, Beijing, and points between. For instance, Lopez found that the Florentine Frescobaldi company imported China silk to London in 1340, perhaps directly from its original source. Given the volume of commerce in the thirteenth and fourteenth centuries, the Frescobaldi company is unlikely to have been the only direct trading link between England and China. One need only recall the Polos' journey to China or the numerous missionaries who traveled to Mongolia and Asia during the thirteenth and fourteenth centuries to recognize another conduit by which information about Asia could reach Europe. As Kroeber notes in his original exposition of the idea, evidence for stimulus diffusion can often be furtive and fragmentary at best. Therefore, it is necessary to examine the context in which ideas and techniques made their way from foreign sources to Christian Europe.[23]

ECONOMIC AND CULTURAL CONTEXT

In what follows, I offer an overview of some of the forms of interaction, especially commerce and conquest, and contact between the Moslem and Christian shores of the Mediterranean with reference to representative scholarship on each issue. By 1100, the world known to Islamic scholars and merchants, extending from the Strait of Gibraltar to the Strait of Malacca, was well connected by trade. Merchants were particularly likely to have been agents of transmission of a new product or idea. They not only traveled widely but also had direct and extended contact with their counterparts from other cultures and were responsible for the selection, purchase, and shipment of foreign goods to their home cities. Goitein stresses "the freedom of communication, the absence of boundaries, the free circulation of money and almost all goods and books." Based on what he judges to be reasonable customs duties, he describes the Mediterranean as a "free-trade community," and he emphasizes the unity of the merchant world. Merchants traveled and settled throughout the Mediterranean world and beyond. In his studies of the Cairo Geniza, Goitein found evidence of "massive population movements under Islam, especially from Iran and Iraq to the Mediterranean area, first as colonists, then as emigrants and even refugees." These movements of people also went from West to East. For instance, he mentions evidence in Egypt of a silversmith from Ceuta (North Africa) fleeing the Almohad terror all the way to Ceylon. The world in which merchants moved was vast, encompassing the entire known world of the time. Their extensive trade networks made the Venetians and Genoese the most widely traveled among Christian Europeans. Not only merchants but also other people traveled widely as well. Despite Heyd's neat categories separating travelers by intended purpose (crusaders, ecclesiastics and knights, pilgrims and merchants, adventurers and manufacturers), Goitein reminds us that "as the Geniza shows, most travelers were also carriers of goods." In addition to such things as trade goods, ideas, images, and techniques traveled great distances as well. Accelerating the whole process, warfare, conquest, and colonization caused boundaries to shift. This was especially true in Spain and Sicily because the gradual reconquest of these lands brought Arabs, Greeks, and Latins together for extended periods of time.[24]

The cost and difficulty of land transport dictated that most long-distance trade in the Mediterranean was conducted primarily by sea. Therefore, ports such as Genoa, Venice, Amalfi, Messina, Pisa, and Syracuse functioned as nodes of communication and commerce connected with other nodes abroad, including Caffa, Tana, Constantinople, Alexandria, Catalonia, Bougie, and London, and extending as far as Peking. Even humble Sienese ports like Talamone and Porto d'Ercole (20 km south of Talamone) were linked to the larger Mediterranean trade networks. A Sienese chronicle records a wealthy citizen's purchase of 120,000 gold florins worth of silk merchandise (but paid in silver, copper, tan bark, and small coins), from a Syrian merchant at Porto d'Ercole in 1338. It is worth noting that the wealthy Italian exchanged raw materials for finished goods, a type of transaction fully consistent with Goitein's characterization of the unequal trade relations that prevailed between Europe and the Muslim World. In imagining the Mediterranean world from the point of view of a people more likely to have flown over it than to have sailed across it, the casual observer tends to exaggerate the distances between its ports or see the water itself as an obstacle to commerce and transport. A modern reader may forget that the "watery lanes" of which Homer sang were still the shipping lanes in the Middle Ages; the sea formed a medium of communication, it was an "inviting sea," not a barrier.[25]

Some travel times will serve as examples of how quickly news and goods could move. Wilhelm Heyd records voyages from Marseilles to Acre, with a good wind, in as little as fifteen days and nights. He cites another journey of eighteen days from Marseilles to Acre, of which only four days were required from Marseilles to Messina. John Pryor finds longer durations: Genoa to Alexandria, twenty-four and thirty-five days at the end of the fourteenth century; Venice to Jaffa, thirty-two days; return trips, which depended upon much rarer easterly winds, took about twice the time. Nonetheless, it appears that Heyd was justified in beginning his discussion of travel in the Mediterranean with the comment, "La *traversée de la Méditerranée* n'offrait pas beaucoup de difficultés."[26]

Modern perceptions of the difficulties of travel to China rely too much on our knowledge of Marco Polo, who remained so long in China that his friends at home initially failed to recognize him. It is easy to confuse the duration of his journey with the actual time it took to travel back and forth to China. However, Lopez's examination of Genoese commercial documents shows that at the beginning of the fourteenth century the trip to China took less than a year by land, about twice that by sea. Furthermore, even the land route from Tana to Cathay, according to Pegolotti, was "*sicurissimo*" day and night. It was this same ease of travel that sped the plague to Europe.[27]

Most Mediterranean ports had *fondacchi* (from the Arabic *fonduk*), buildings in which the goods of foreign merchants were kept under the watchful eyes of the local authorities. Although foreign merchants were often housed in a sharply delineated quarter, treaties frequently accorded them liberty of their host city. In the *Decameron,* Boccaccio describes a fondaco in Palermo:

> In all the coastal cities which have a suitable seaport, it was common practice, and perhaps still is today, for a merchant arriving there with cargo to unload it and then transport it to a warehouse (*fondaco*), which in many places is called a customhouse (*dogana*) and is under the control of the commune or the ruler of the place, and there, after those in charge have been supplied with a written list of all the merchandise and its value, they provide the merchant with a storehouse (*magazzino*) in which his goods are placed under lock and key; and then the customs officers (*doganieri*) record in their account book (*libro della dogana a ragione*) all the goods belonging to the merchant, and later on, as he withdraws his goods, all or in part, he is made to pay the required duties.[28]

In the local bazaars, with their hundreds of stalls, what novelties awaited the foreign merchant? Would he desire a sherbet, a demitasse of coffee, a jar of candy, even a soda?[29] Pegolotti lists hundreds of goods the Italian merchant should be on the lookout for, along with criteria for selecting the best. A network of traders reaching to the East Indies and China brought many of the well-known "minute" spices (the commodities "sold in small quantities at high prices," now commonly thought of as spices) such as cinnamon, fenugreek, and peppercorns to wholesale markets in the Levant and Egypt. Other, bulkier "spices," which would now be categorized as industrial, medical, or other chemical commodities like sugar, various grades of alum, or gum Arabic, also came great distances. Together they formed the basis for an evolving global trade that brought Christian Europeans to foreign ports for extended stays, exposing them to new forms of luxury, pastimes, and behavior. Scholars are still working out the details of how living in close contact with foreign customs and goods affected the merchants and other travelers abroad. For instance, it is unknown whether Europeans acquired a new taste for bathing in foreign lands when they arrived in foreign ports or if they were already conditioned by a tradition of public bathing, unbroken since Roman times, to seek the baths near their new quarters. Documents like the treaty of 1234 allowing the Pisan merchant community at Bougie access to the public baths once a week are silent as to whose idea it was for the Pisans to bathe regularly. If behavior such as bathing is hard to unravel, even though it is reflected in the documents and archaeological remains, imagine how much more difficult it is to measure the impact of ephemeral activities such as music, dance, or storytelling.[30]

Difficult or not, literary historians have tried to assess the impact of storytelling upon the written word. Menocal is a representative of a dissident tradition of scholarship that asks how many of the literary motifs in the *Decameron* were transmitted orally in the multi-lingual communities of Spain or Sicily. Did Italian merchants hear stories from other merchants, Christian, Muslim, or Jewish, as they passed their months abroad, waiting for a caravan to arrive or for a flotilla to assemble for the journey home? She raises the possibility that Boccaccio perhaps heard Arabic stories from Italian or foreign merchants in Florence or Naples. Menocal cites other scholars who ask if Dante borrowed the architecture of the *Inferno* from an Arabic source. Likewise, one could also ask the analogous question regarding the extent to which Arabic literature incorporated European influences. Until a broader cross-section of scholars recognizes the degree to which Christian, Jewish, and Muslim merchants came into contact with each other, such probable scenarios will unfortunately be discounted, preventing the formation of a more accurate image of the Middle Ages.[31]

Economists have long understood the part "import substitution" plays in the transformation of a consumer of goods into a manufacturer. One way in which a consumer such as an Italian city-state can meet a rising demand for an imported good, and thus redress an imbalance of trade, is to substitute its own manufacture of that commodity. What is not yet known are the details of the processes by which those bales of buckram, cotton, chiffon, damascene, fustian, gauze, moiré, mohair, muslin, percale, sashes, satin, sequins, silk, tabby, and taffeta were supplanted by domestic textile production in Florence, Bologna, Lille, and dozens of other cities. Only a few of the more important medieval textile industries and fibers such as wool, silk, and cotton have found their historians.[32]

Ports were gateways to ideas as well as commodities. White suggests that the port of Amalfi was the pathway by which the ogival arch made its way from that city's trading partner, Fatamid Egypt, to Montecassino by 1071. A decade later, Abbot Hugh of Cluny noticed it at Montecassino and adopted it into the new architecture that was evolving out of the Romanesque in France. From there it entered the mainstream of what would later be called the Gothic style. (Although entirely plausible and adequately documented, it is difficult to imagine that such a ubiquitous feature of the Islamic world had only a single route of entry to Christian Europe.) Even so, few innovations can be traced so completely nor should we expect to do so for most of them.[33]

White's query regarding this one architectural feature leads us to ask what other sorts of architecture, urban amenities, techniques, or innovations might a traveler abroad have seen and desired to bring home? The huge scale and concentrated populations of Islamic cities required innovations that would not be necessary in Europe for several centuries, such as the numerous floating mills moored to the banks of the Tigris and Euphrates that ground flour for Baghdad's burgeoning population. In his critique of Singer, White trumpets the substitution of power for the sweat of laborers as one of the signal achievements uniquely characteristic of "Western" ideology and technology, calling it "the novel concept of labor-saving power technology which is one of the pillars of the modern world." We can only speculate as to how better knowledge of the application of power to industrial and craft processes in the Islamic world might have altered White's generalization about the uniqueness of Western ideology. White acknowledges the prestige of medieval Islam, saying "Samarkand and Cordova, Baghdad, Damascus and Cairo were magnificent centers of art and thought as well as commerce." But then he asks, "what original ideas and methods did they provide in technology?" By posing this question, he shows that he has overlooked the technological underpinnings of their scale and wealth, and the ways in which Islamic engineers and their patrons pioneered some of the very techniques that would come to characterize western technology for him.[34]

White also discusses the consequences of the application of power to agriculture and attributes the "revolutionary productivity" of the new agriculture to the improved harness and the three-wheeled heavy plow. Unfortunately, he ignores the introduction of the very crops that would come to be so intimately linked with southern Europe—rice, sugar, new varieties of hard wheat, eggplant, and citrus fruits—all of which came to Europe through Islamic neighbors and trading partners or through the reconquest of the Iberian peninsula. Until the introduction of crops from the New World, species transmitted through the Islamic world made "renaissance" agriculture and cuisine different from their Roman antecedents. More importantly, he fails to assess the crops' impact upon productivity, a task that has yet to be undertaken. The whole issue of scale and the impact of increasing the scale of an operation or process deserve more attention in the late-medieval and early-modern eras. The need to increase the scale of a process, whether it be the exploitation of a natural resource, the production of a commodity, or the disposal of urban waste, is one of the defining functions of the modern engineer. The task of enlarging or replicating a process provides exactly the milieu in which engineers flourish, and we can expect their numbers to increase whenever a patron requires this sort of task to be performed, especially if the patron is an institution or a rich individual able to hire more than one person with sophisticated engineering skills at a time. Two crops in particular provided precisely this stimulus: sugar and paper used large amounts of power and large teams of skilled and semi-skilled labor.[35]

An examination of sources documenting trade, exploration, and warfare in eastern cities reveals that they usually contained certain strategic necessities. For instance, besides having industrial installations such as Baghdad's floating mills and the wind mills that ground grain along the silk road in Seistan, eastern cities boasted trip hammers that hulled rice, pounded gold and other ores, reduced rags into pulp, or fulled cloth. Norias in Spain and Syria lifted water day and night unattended. As with the ancient Roman world, the degree to which urban and industrial enterprises relied on water power in the Islamic world has been vastly underestimated; thus, inferences suggesting that the Muslim societies lacked the desire to employ labor-saving devices can no longer be accepted as completely as they were before.

Another area in which the application of power allowed technological advances was mining. White criticizes Singer for his assertion that European mining had not advanced since Roman times but assumes these advances were autochthonous. But the mobility of miners made it likely that states would import foreign techniques when they hired miners from outside their territory to exploit local resources. Did the techniques European merchants and entrepreneurs observed, such as those guided by Pegolotti to construct the alum mines of Phocea (now Eskidje-Fodja), contribute to the post-classical innovations in mining first documented by Agricola in his *De re metallica?* Hassan and Hill, who devote an entire chapter to mining in their study of Islamic technology, say that the research to answer such questions has yet to begin.[36]

In addition to industrial applications of power, travelers observed luxury goods and processes as well as certain culinary novelties such as the two thousand points from which ice water was distributed in tenth-century Samarkand. Those Europeans who visited Islamic gardens found them filled with fragrant plants like lemons, limes, sour oranges, spinach, watermelons, artichokes, and the crocus from which saffron is gathered. What began as specimen plants in study collections, novel vegetables or grains, or oddities intended to impress visitors ended up transforming European agriculture to an extent that has only been evaluated in Spain.[37]

Amidst these plants, collected from around the known world, stood fountains that played water tricks upon the unwary visitor. Medieval water tricks first appear in Islamic gardens and in the books of Islamic engineers such as the *Book of Ingenious Devices* of the Banu Musa (ninth century C.E.) and *A Compendium of the Theory and Practice of the Mechanical Arts* of al-Jazari (twelfth century C.E.). Later we find them in the private gardens of Naples, the Veneto, and Tuscany. Waterclocks signaling the hours by means of automata may seem innocent of practical application at first glance, but they contain the seeds for advanced technologies of automation and control. It is likely that some of these devices originated in the writings of Hellenistic engineers such as Ctesibius (third century B.C.E.) and Hero of Alexandria (first century C.E.), but it is not always easy to trace specific devices based on the verbal descriptions that survive.[38]

Exposure to an exotic product, be it an eggplant, the game of chess, an astrolabe, perfume, incense, a silken sash, or a translation of Aristotle, could awaken the desire to take it home. Some of these things like manuscripts, chess pieces, or bolts of silk could easily be brought back in one's personal luggage. Larger technological processes, like growing rice, sugar production, or sericulture, constituted entire systems comprising many parts knit together by highly skilled labor, suitable markets, and an accepting cultural climate. They could require years to transport all the components and to obtain the skilled operators from zones where these crops and products were well established, let alone to develop a market for the product. The degree of success with which products or processes were transferred ran the entire gamut from the fragmentary stimulus diffusion to government sponsored transfers of entire technological processes. Governments and rulers tried to protect some of the processes from export, especially those that produced profitable monopolies like sericulture, papermaking, and glassblowing. Rulers took steps to ensure that foreigners could not observe critical processes and to prevent skilled workers from migrating. Perhaps it was the necessarily furtive means of acquisition, using techniques that today would come under the heading of "industrial espionage," that makes retracing the routes of transmission so difficult. There seem to be more good stories than verifiable accounts. In rare instances, we actually know how a complex technological process, including the essential skilled technicians, was transferred. Such examples include the treaty of 1277 with Aleppo, by which Venice obtained the secrets of Syrian glassmaking (secrets Venice would protect by isolating its glassmakers on the remote island of Murano in the Venetian Lagoon), and the capture of Chinese papermakers at the Battle of the Talas River and their transfer to Samarkand in 751.[39]

More typical of the state of our knowledge is our inability to identify crucial steps of paper's transmission to the West. For instance, we still do not know the stages by which mechanized papermaking reached Fabriano by the 1240s or how to explain the appearance there for the first time of three important innovations: the use of water-powered, studded hammers, the addition of gelatin sizing to the pulp, and the incorporation of watermarks by the 1280s. One telling detail gives us a clue: there is evidence of paper production in the countryside near the port of Genoa, based on the Spanish model in 1210, about thirty years prior to Fabriano. This makes Genoa a likely stepping stone to Fabriano. Why Fabriano? Because in addition to its proximity to clean, fresh water, and for both the paper-making process and to power its rag hammers, it was in a region already producing flax for linen, a product of Jativa, as well. Furthermore, it was far enough from Genoa to be able to supply a different market.[40]

Frequently the attempt to follow a particular innovation's route to Italy leads to a port where the trail is then lost. Such is the case with so many instruments of trade: the letter of exchange, the letter of credit, tariffs, gabelles, customs, and maritime insurance. Were the Genoese and Venetians so inefficient and so unobservant of their Muslim trading partners' practices that they needed repeatedly to re-invent each of these devices, or did they acquire some of them from their commercial brethren? A suggestive story about how the process might have worked can be found in the life of Leonardo of Pisa (ca. 1170- after 1240). His father, Guilielmo, while in charge of the Pisan trading colony in Bougia (now Bejaia, Algeria), sent his son to Arab masters to learn business skills. When Leonardo, also known as Fibonacci, returned to Pisa, he published these revolutionary techniques of place notation and "the new Indian numerals" in his *Liber Abbaci* (1202). Could other business techniques and instruments have been transmitted in similar ways to the sons of merchants living abroad, who then introduced the techniques into their family businesses? In many cases, diffusion among merchants provides a more likely scenario than independent invention.[41]

While merchants, pilgrims, warriors, and bankers were bringing commodities and manufactured goods back to Italy, thus stimulating demand for those goods and ultimately the desire to produce them domestically, scholars and their patrons were involved in another sort of exchange. The Iberian Peninsula was famous for its schools of translation, as the reconquest encompassed established centers of learning including Toledo, Ripoll, and Córdoba. Southern Italy was also host to centers of translation, especially during the reigns of Roger II and Frederick II, the rulers of the trilingual Hohenstaufen courts in the Two Sicilies. The results included not only the transmission of the entire Greek corpus of science and philosophy, but also the introduction of significant Islamic additions that shaped the ways in which westerners were to interpret Aristotle, Plato, Philo, and Heron, which together formed much of the basis for what the newly founded universities would teach for centuries to come.[42]

From the bottom of the social strata came an unexpected source of exchange of ideas with the East: slaves. Until recently, slaves have been invisible in the literature on medieval Tuscany, leading scholars to overlook them as a means of contact with the east. Historians abandoned this assumption when Giulio Prunai and Iris Origo documented the importation of hundreds of slaves to the region, conclusively demonstrating that the institution was widespread in medieval Tuscany. White was the first historian of technology to build on this work, suggesting that the word "Tartar" in Taccola's notebooks could have referred to Tartar slaves. From documents published by Prunai we know that official Sienese records used the term Tartar to denote the specific place of origin for some of Siena's many slaves during Taccola's career. The details of what skills these slaves possessed or ideas they actually contributed will likely never be worked out, but their importance as a potential means of diffusion cannot be overlooked.[43]

It is hoped that by synthesizing the research of historians of individual technologies, such as windmills, watermills, mining, or textiles, with the data being gleaned by economic historians about the nature of commercial relationships in the Middle Ages the impasse in research regarding the diffusion of technologies across cultural frontiers in the Mediterranean can finally be breached. Once the larger model of ongoing communication throughout the known world, long recognized by Lopez and others, is adopted by other historians, the gaps in Needham's chronological table (reproduced in Singer) can be narrowed and the hypotheses regarding steps along the routes of transmission can be tested and perhaps retraced. This sort of scholarship cannot happen until a program of publishing Arabic technical manuscripts is underway and an equally ambitious program of archaeological excavations in the medieval ports from London to ports in China has begun to yield data. Until then, we are forced, like White, to use a model like Goitein's of a united merchant community to interpret the available artifactual evidence. From this model we can only make bold inferences as to the nature of medieval engineering and its sources and perhaps indicate fruitful paths for future research. Although this chapter has focused on European borrowings from more technologically advanced cultures, a deeper investigation will reveal what military historians already know about the period under study, that techniques moved both ways across cultural frontiers.

ANTECEDENTS

With these difficulties in mind, we will turn to the physical evidence, beginning with potential antecedents to the Sienese aqueducts. In 1940, White made the sweeping claim that there was "no evidence of a break in continuity of technological development following the decline of the Western Roman Empire." The archaeological evidence provided by the Sienese aqueducts, or *bottini,* reveals vestiges of ancient skills in surveying, digging, and channeling water to where it was wanted. The techniques required to drain a mine and to fill a fountain with water are identical; only the goals differ. Civilizations rose and fell in Central Italy, but the core complex of technological skills that underlay them continued largely unbroken in a few institutional settings such as monasteries and urban settlements. The skills required to mine metals, quarry stone, alter the landscape through drainage, and direct water were known in varying degrees to every civilization inhabiting rural Tuscany for the past three millennia. In their ability to modify their environment, providing a year-round supply of water in a land of highly seasonal rainfall, the medieval Sienese were no exception.[44]

Whereas the local roots of mining, water supply, and building technology in Tuscany can be traced to the Etruscans, we do not know what additions came through the later two waves of diffusion—the Roman and the Islamic. The Etruscans enjoyed a reputation among their Roman conquerors for having a deep understanding of the mysteries of hydrology. The archaeologist Jacques Heurgon writes, "Etruria was where the best water-diviners were found, and these were not common water diviners . . . [;] they knew for example how to find lakes of subterranean water by examining peculiarities of vegetation and how to bore . . . artesian wells." Glick suggests, however, "that deep-drilling techniques also had an Islamic diffusion, to judge by the interval between its mention by al-Biruni, writing in Damascus in the early eleventh century, and its first European documentation, at Artois in 1176." Heurgon points to the first wave of diffusion, the pre-Roman wave, and Glick points to the third, the Islamic wave of diffusion.[45]

In the field of agriculture, the Etruscans used analogous skills to drain land. They reclaimed swampy fields by cutting huge networks of drainage canals through the hard volcanic rock called "tuff." These shafts drained wetlands into neighboring valleys, not only opening new farmland and pastures but also eliminating a habitat conducive to malaria. Finally, and most germane to the Sienese experience, they built fortified, hilltop cities such as Fiesole and Perugia and supplied them with water.

Some of the most signal achievements of Roman engineering, such as the Cloaca Maxima, have been attributed to the Romans' Etruscan forebears. The professor of water-supply engineering and public works at the École Spéciale des Travaux Publics, Jacques Bonnin, emphasizes the Romans' debt to their predecessors but says nothing of the sources upon which the Etruscans may have drawn. The distinguished mining engineer, Henri Goblot, whose extensive work in the Middle East led him to write a history of qanats, suggests that Etruscan cuniculi fell into disuse under Roman rule because at the time the Romans took over the Etruscan countryside, they lacked the understanding of hydrology necessary to anticipate the consequences of their neglect, as when, for instance, the Romans destroyed the cuniculi at Veii in 396 B.C.E. Furthermore, he asserts that the Romans did not have the social organization outside the city necessary to maintain the cuniculi had they understood the need to do so. Goblot observes that under Roman rule, the countryside around Rome, "so fertile and so rich in Etruscan times . . . became a sad region, desolate and infested with malaria" until the 1930s. Goblot writes that whereas the Etruscans deeply understood the various means of manipulating the landscape and the precise effects their drainage systems had upon the countryside, the Romans failed to comprehend the ways in which drainage shaped the rural environment. He calls the Romans "completely ignorant" of such subtleties. Heurgon points out that "a drainage canal, if it is not looked after, becomes just as dangerous a place as a marsh." Heurgon asserts that by the time the Roman conquerors could effectively command such a large undertaking as the maintenance of the drainage tunnels, the once-drained areas had already become regions infested with malaria.[46]

As to how the Etruscans learned the technique of drainage and whether the cuniculi were an independent invention or a diffusion of the qanats of Iran, Goblot offers no firm answers. He speculates that if the Etruscans did migrate from the East, they may have learned mining and drainage techniques from the peoples inhabiting the Iranian plateau, who originally developed the adit and the qanat. He sees the knowledge of qanat construction as a natural outgrowth of experience in draining mines, of which the Etruscans had many in Tuscany and on Elba. He concedes, however, that the question has yet to be answered, and until then he classifies cuniculi as an invention independent of qanats.[47]

Millennia of human occupation along the hills of Rome have obscured most evidence of pre-Roman building programs. Therefore, the Etruscans' legacy is most evident outside the Eternal City. Like the tribes that settled Rome, the Etruscans preferred defensible hilltop sites, above the malarial lowlands, for their cities. Thus they encountered the same difficulties in obtaining water as did later, medieval inhabitants of the same sites. For instance, the wells in Perugia, an ancient hill-town one hundred kilometers to the southeast of Siena, demonstrate one aspect of the Etruscans' engineering abilities: well-digging. The Pozzo della Piazza (or "Etruscan Well," as it is also known), is around 3.3 meters in diameter and extends to a depth of more than forty-seven meters beneath the present Piazza IV Novembre. Lined with tightly fitted cylindrical sections of cut stone, joined without mortar, it supplied a siege-proof source of water to the Etruscan settlement. Structurally sound after more than two thousand years, the well in Perugia continues to testify to its builders' skills in hydrology, excavation, and stonework.[48]

Whether the Etruscans' facility in digging deep shafts and drainage canals—even through hard volcanic rock—was enhanced as a result of experience gained in the operation of their mining enterprises, or vice-versa, is not possible to answer. But it is clear that these two activities together expanded the range of skills upon which excavation projects could draw. The Etruscan economy in Tuscany depended upon the ability to mine, refine, and use ore to produce metal objects such as bronze weapon-points and fibulae. The Etruscans traded such items throughout the Italian peninsula and beyond. Besides minerals, they mined and carved alabaster for local use and for trade. They also quarried stone for construction, using it not only for wells, but also for their famous necropoleis and the funerary decorations they housed. Here it is appropriate to recall that one of the most frequent obstacles to overcome in hard-rock mining is the evacuation of water. In the course of surmounting the challenge of draining mines, the Etruscans accumulated a complex of skills applicable in other areas. Agriculture and urban water supply were the most important applications outside of mining proper.

Although they allowed the rural drainage works in the conquered territories to fall into disuse, the Romans who succeeded the Etruscans did develop skills in the conveyance of water to cities. The most famous of their projects are the aqueduct systems by which Rome and many other cities within the empire were supplied with water. The eleven aqueducts bringing water to Rome totaled 502 kilometers in length. One aqueduct in Rome, the Aqua Marcia, conveyed water from ninety-one kilometers outside the city; another, the Anio Novus, was eighty-seven kilometers long. Contrary to popular impressions, most of the aqueducts were subterranean and not elevated until they crossed the valleys on the outskirts of Rome. Though perhaps reflecting a less subtle understanding of hydrology than their Etruscan predecessors, the Roman aqueducts left imposing physical evidence of their builders' engineering achievements that was to exert a powerful influence on their late-medieval and early-modern successors. The rediscovery and diffusion of ancient texts, especially those by Vitruvius and Frontinus, coupled with the excavation of Roman artifacts and the restoration of some of Rome's aqueducts, increased late-medieval knowledge of Roman practice. Even the Sienese, occupying a site that lacked ancient archaeological remains, felt the weight of history, perhaps even more so than those cities, such as Florence, that could boast of Roman ruins. The Sienese responded to the absence of physical ruins by inventing a Roman past. They adopted the image of the she-wolf suckling the twins Romulus and Remus as the city's symbol. The Sienese embraced the myth that the city was founded by Senius, son of Remus, and explained that the design for the city's black and white emblem, the Balzana, was derived from the smoke of Senius' sacrifices to Diana. They believed that their Duomo was built on the site of a temple to Minerva and that the Campo was built on the site of a Roman forum. Noble Sienese families such as the

Piccolomini claimed descent from old Roman families. The Tolomei claimed the Ptolemies as ancestors. All this attention to real and imagined Roman precedents—and the ensuing battles between "ancients" and "moderns"—has served only to draw attention away from the greatest sources of innovation, which came from abroad.[49]

PARALLELS

It is important to look at similar systems of water supply because they are the most likely source of innovation in Siena. Contemporary parallels to Siena's aqueduct system can be found within and beyond Europe. One can find similar water systems on the Italian peninsula, in Northern European towns from England to Bohemia, and in many alluvial plains in lands stretching from the Iberian Peninsula to the Indian sub-continent. In general, the farther the example is from Tuscany, the less it resembles the system in Siena, either in its physical details or the social structures that shaped its construction and maintenance. The examples that follow begin with those sites nearest Siena.

The hill town of Colle di Val d'Elsa, about twenty kilometers northwest of Siena, has a system of bottini very similar to those of Siena. Like Siena, the medieval core of Colle is built atop a ridge, formed by the remains of various sedimentary strata, including tufa. The tufa under Colle may once have been geologically contiguous to the tufa upon which Siena is built. The system of bottini in Colle, though known to local geologists and archaeologists, remains unpublished. One main bottino extends the length of the old city, Colle Alta, under the principal thoroughfare. Its total length is less than three kilometers.[50]

Orvieto, a hill town about one hundred kilometers south of Siena, and a little smaller in size, has a system of medieval aqueducts. A recent study of the system shows Orvieto's water supply to have had Etruscan and Roman foundations, originally consisting of wells and cisterns. The system of aqueducts that replaced the earlier wells dates from around 1276. The water feeds a total of a dozen fountains, most dating from the thirteenth and fourteenth centuries. The main aqueduct brings water from two springs about three kilometers southwest of the city. After constructing its aqueduct, Orvieto sank a well around sixty meters deep. The regulations governing the use of water were similar to those of Siena.[51]

Bologna, 185 kilometers north of Siena, has an underground aqueduct dating from around 25–15 B.C.E., during the reign of Augustus. Its source is the Setta River, at a point right before it flows into the Reno, south-southwest of Bologna. From the Setta it brings water eighteen kilometers underground to the oldest section of the city, the original Roman grid. The cross section of the tunnel is generally rectangular, sometimes reinforced with vaulted masonry roofing. Like that of so many other aqueduct tunnels, the cross section is just big enough for a worker to pass through, measuring about .65 meters wide by 1.9 meters high. The slope along its length, like the main bottino for the Fonte Gaia in Siena, is a little less than 1:1000. The aqueduct fell into disrepair around the sixth century and was not rediscovered until the fourteenth century. Only in the nineteenth century did the city restore it to operating condition.[52]

A water supply system in Exeter, England, used tunnels carved through rock of about the same size as those in Siena, but they functioned quite differently. Instead of having a small canal trenched along the bottom, the Exeter tunnels provided an accessible conduit through which ran a piped water supply. Exeter's underground passages run under High Street and supply the town with spring water. The system dates from the thirteenth and fourteenth centuries. A summary prepared for the general public notes that "The oldest part that can be visited today was constructed in 1346–49 for the Cathedral's water supply." The city built its own system during the years 1420–1429 and operated it until 1832, when it was closed to curtail a cholera epidemic. The underground passages were only rediscovered in the twentieth century.[53]

The water system at Canterbury instead used pipes directly buried in the soil. It was built in the twelfth century to supply the chapter house of the cathedral. Because the pipes were hidden from view, the builders drew a diagram to facilitate maintenance. Thanks to that diagram, which survives from the twelfth century, we know the system's extent. Both Exeter and Canterbury were ecclesiastical systems. Their similarity supports J. B. Ward-Perkins's and Paolo Squatriti's observations that many skills in water management owe their survival to religious institutions' filling the gap left by the disappearance of Roman civil authorities. It thus appears that religious institutions supervised their own and sometimes a neighboring town's water supply in the early Middle Ages in at least one region north of the Alps. The institutional stability and continuity allowed the survival of technologies and skills even as governments collapsed and the population of urban settlements fell.[54]

Two ancient forms of water supply originating outside Europe deserve special mention here because of the striking similarities they bear to the two types of Sienese aqueducts. The first is the well-documented qanat. The second, a related form, the geographic distribution of which overlaps that of the qanat, is called a spring flow tunnel. Zvi Y. D. Ron identified the spring flow tunnel as a distinct form, and in 1966 he became the first person to publish a description. He followed his initial publication with two more articles, in 1985 and 1986. His archaeological research has uncovered over 250 of these tunnels in Samaria and Judea, of which he has measured and mapped over a hundred. Ron's work offers a major new analytic framework for examining the nature of the oldest Sienese aqueducts.[55]

Both qanats and spring flow tunnels have their origins in the Middle East almost three thousand years ago, and both are still in use. Ron defines their functional similarity as follows: "Both are excavated tunnels which branch out into the saturated aquifer to extract water by gravity flow in quantities sufficient for irrigation."[56] Both have been used concurrently in the hilly regions of the Holy Land since almost the second millennium B.C.E. The dimensions of each in cross-section can be quite similar, but here the resemblance between the two ends. According to Ron, their differences lie in their geographic distribution, geomorphology, origins, agricultural determinants, the patterns by which the tunnels are developed, the role of the access shafts, the incline of their water channels, techniques of construction, amount of maintenance required, and the patterns of ownership. Spring flow tunnels take a more organic, irregular form as a result of their natural origin, lack of plan, and continual extension. Spring flow tunnels are the result of geologic processes that began millions of years ago. The excavations that give them

Figure 5.1 Fonte al Pino around 1900. The exterior, much simpler than many of the other fountains gives little hint of how different the interior of this spring flow fountain is. From Bargagli-Petrucci, *Fonti di Siena.*

their form are based on observation, not on a plan. The need for enlargement is a result of the tunnels' effect upon the hydrology, usually one of lowering the water table. Each time the water table sinks below the tunnel bottom, the excavations must resume in order to intersect with the lower water table. Qanats appear to represent a new level in sophistication in the techniques of exploiting groundwater resources. Instead of being the result of empirical searching for more water near a spring's outlet, qanats are built all at once, according to a carefully surveyed plan, though the process can take several years. The plans are based on finding the most direct route between the water-bearing rock and the village or irrigation project that will be supplied with the water. Their planned development represents a new level of hydrological knowledge gained over perhaps centuries of trial and error. The new knowledge allows the *muqannis* (qanat builders) to read a landscape's hydrology in order to plan the course of a qanat. As a result, qanats are longer, more uniform, and the surface above them is marked by periodic ventilation shafts.

Because of their greater length, qanats are more expensive to build than spring flow tunnels. Unlike spring flow tunnels, qanats do not deliver a drop of water until they are completely excavated. Spring flow tunnels typically exploit the very springs around which villages were first built. Having often preceded a village's foundation, the springs and ensuing spring flow tunnels are publicly owned. Qanats are usually privately built and owned, the result of entrepreneurial initiative taken to create new arable land in the desert. The external source of financial backing is likely to have contributed to the development of muqannis as a distinct specialized occupation and the scale and number of such excavations to the regularization of the product.

The reader will recall that Siena has two types of bottini, the older, shorter, meandering type that feeds the fountains built before the Fonte Gaia, and the newer, much longer, more regular type represented by the Fonte Gaia system. A comparison of the Sienese bottini with other systems reveals that the older bottini are quite similar to the spring flow tunnels documented by Ron. The site in Siena that most closely resembles typical urban spring flow tunnels that Ron surveyed is the recently discovered Fonte delle Monache, located below a medieval convent in the northeast corner of Siena. Like the fountains Ron detailed near Jerusalem, the Fonte delle Monache, or Fountain of the Nuns, is a completely enclosed fountain complex about seven to ten meters below the sloping surface of the ground. It can be reached by either of two staircases: one from an adjacent garden, the other from the cloister's courtyard above it. The stairs, like the walls, ceilings, and basins, are completely carved from the tufa. The author's visit to the Fonte delle Monache in the summer of 1996 confirmed the striking resemblance it bears to the spring flow tunnels Ron identifies. Likewise, the oldest fountains in Siena, including the Fontanella, the Fonte Branda, the Fonte Nuova, the Fonte Ovile, the Fonte Peschaia, and the Fonte Follonica, also resemble spring flow tunnels. The flow tunnels of the Holy Land and those of Siena resemble each other sufficiently to categorize the older bottini of Siena properly as spring flow tunnels. The only notable difference between the spring flow tunnels of Siena and those in the Holy Land is that the latter are cut out of harder rock than the Sienese tufa, which did not leach so many calcites into the water, and thus required less maintenance.

The newer, longer tunnels that bring water from a distant source, like the Fonte Gaia aqueduct, resemble what Ron, Goblot, and others define as qanats. In fact, not only does the Fonte Gaia aqueduct—also called the Bottino Maestro—bear a physical resemblance to qanats, but its construction (all at once) and the way in which it has been subsequently expanded (by a pattern of branching to other springs) also resemble the means by which qanats are constructed more than they resemble Siena's spring flow tunnels.

Figure 5.2 Fonte al Pino Interior around 1900. The shape of the interior shows how the fountain was hewn from the rock itself. The lack of decoration make the similarities to ancient fountains in the Near East more apparent. From Bargagli-Petrucci, *Fonti di Siena.*

Ultimately, Ron's work may represent the first phase of an expanding literature of pre-industrial, gravity-fed water supply. This literature will interest historians seeking a wider context in which to understand Roman aqueducts, medieval aqueducts, and qanats, each of which has until recently been studied in relative isolation from others. Such comparative historical studies will also perhaps interest contemporary water supply engineers and policy makers attempting to design ecologically sustainable systems for the world's arid regions. The success over millennia of past systems may find useful application as models in the future. Technologists may find the traditional systems described here especially appropriate to those towns and villages in the developing world that are rich in human capital and ingenuity but without access to the financial resources necessary to build and maintain high-tech, energy-intensive water supply systems. Conservation, by using water more than once, could also be the most cost-effective means of expanding the net water supply of cities in the desert, like Los Angeles or Las Vegas.[57]

Unfortunately, until proper archaeological investigations of the Sienese aqueducts of both sorts are carried out, the question of diffusion versus independent invention cannot be settled. Without more evidence we cannot be any more certain about the origins of the Sienese aqueducts than Henri Goblot was thirty years ago when he declined to speculate on this point. Nonetheless, the work of scholars like Goblot, Ron, and Roberta Magnusson, allow us, for the first time, to place the Sienese bottini in a context broader than the several Italian hill towns with similar systems. Having addressed the question of uniqueness we can now turn our attention to the cultural consequences of the long-term investment medieval Siena made in its infrastructure and the effect that ongoing public works projects had in raising the level of technical expertise within the city.

Chapter Six

Sienese Engineers and Their Urban Patrons

An examination of the careers of three successive generations of Sienese engineers—Taccola, Francesco, and Biringuccio—suggests that the occupation of engineer developed in ways that parallel that of artists and architects. What they all had in common in Siena was to heighten the role of urban patrons, often city governments themselves. Engineers and city governors often shared an ethos of pragmatism, efficiency, and gradual, necessity-driven innovation regarding the ways in which their work should be done. For instance, when William Bowsky wrote, "their style was, if anything, a pragmatic one that emphasized the institutionalization, formalization, and regularization of practices and offices that worked. The aim was efficiency and effectiveness, successful functioning," he was describing an ethos from the fourteenth century that would not be unfamiliar to an engineer in the twentieth century. However, he was not writing of engineers in the fourteenth century; he was attempting to explain why the members of the Nine, Siena's ruling oligarchy, who created the context in which engineers and architects were to flourish, enjoyed a longevity of rule "surpassed only by the Venetians." The values and goals Siena's rulers shared with the emerging occupation of engineering made Siena a good place for engineers to flourish. Moreover, Siena's location at an economic and cultural crossroads between East and West gave the city's engineers an exceptional toolkit of technical processes and solutions upon which they could draw in their service to the city. The conjunction of intellectual raw material and cultural climate allowed engineering to flourish in Siena out of proportion to its size and power.[1]

In 1972 Lynn White, Jr. argued that, "from the 1440s until the 1480s, Siena, which was best represented by Mariano di Jacopo detto il Taccola and by Francesco di Giorgio, seems to have been the chief source of [European technological] innovation." Although the details remain the subject of ongoing research and debate, the case he made for the impact of Italian engineers upon the rest of Europe can be justified, at least from a quantitative point of view. Gustina Scaglia noted in 1992 that Francesco's drawings, which embody many of Taccola's innovations, have been copied "more times than those of any artist of the fifteenth and sixteenth centuries" and supply many of the core technologies represented in the genre of printed literature called "Theaters of Machines." What remains to be done is to explain "Why Siena?" Scholarship since 1972 has gone a long way toward providing parts of the explanation. Publications by historians of Siena who are not specialists in the history of technology have provided not only valuable new information, but more importantly a new way of looking at what makes Siena different from other cities in Italy. Until recently, and especially among Sienese scholars, there was a tendency to let the city's beauty and charm, in the twentieth century, obscure those factors which might have been the causes of differences. Wolfgang Braunfels, Bowsky, Titus Burckhardt, Judith Hook, and Richard Goldthwaite have by stages allowed us to see more clearly what those differences are.[2]

Although most of the relevant sources had been published by 1910, it was not until 1953 that Braunfels placed Siena's 1297 design restrictions in a wider context, arguing that they constitute the oldest known building code in Europe. The use of hundreds of citizens from many ranks of Sienese society to do much of the Commune's day-to-day administrative work appears to constitute a peculiarly Tuscan practice. Smaller towns, like Siena, recruited an even higher percentage of its citizens to serve in various capacities than larger towns. Roles for citizens included committees to select sites, choose artists, and settle minor disputes. The city also hired many consultants when an issue required a higher level of expertise. In addition to the officially appointed members and paid consultants, the city provided monetary incentives for its citizens to report violations of its ordinances to the authorities. Together, these various measures not only extended the reach of the city government well beyond what its paid staff could do themselves, but it also had the consequence of making an unusually high proportion of its citizens directly involved in shaping the city and personally invested in the outcomes of design decisions.[3]

One of the scholars most knowledgeable about the day-to-day workings of Sienese government is Bowsky, who immersed himself in the Sienese archives for several decades. As one would expect, Bowsky's work is the most nuanced explanation to date of the inner workings of the Sienese administrative apparatus, and thus his work provides the best place to look for tangible differences between Siena and its neighbors. He focused most of his attention on the documents left behind by the fiscal and legislative organs of Siena's government. His examination of the decision-making process from the thirteenth through the fifteenth centuries led him to conclude that, indeed, Siena's government and its priorities were peculiarly Sienese. Bowsky contends that the government of the Nine, which ruled during most of Siena's Golden Age, set a tone for artistic and cultural production that was unique among Italian city-states. He argues that the city reserved its highest rewards for those who contributed to the public good, rather than individual patrons. He further asserts that much of the difference between the Sienese and Florentine cultural heritages was not based on size, wealth, or geography, but can be explained by this reward system. I argue that the cultural climate the Nine created also benefitted masters in the building trades and the emerging occupation of engineers because they produced those goods and services most valued in Siena: public works. Bowsky's thesis—based on his intimate knowledge of the workings of the longest-lasting Sienese government—could help explain the emergence of the "Sienese School of Engineering" that White identifies.[4]

The construction and maintenance of urban systems such as Siena's provided an important training ground for engineers before the skill set for an "engineer" had been completely defined. In this way urban water supply provided a cultural context similar to that fostered by the entrepreneurial construction of qanats in the Islamic world. Like the architects with whose skills and tasks they overlapped, engineers in the thirteenth and fourteenth centuries had no guild or university examination or other ready means of defining the boundaries to their occupation.[5] James Ackerman describes their training as informal at best:

> To this preparation for the practice of architecture—best a haphazard one—was added a kind of schooling that the grandiose projects of the period made available to most of the potential architects in the urban centers. Projects such as the construction of St Peter's and the Vatican Palace brought innumerable artists and artisans into close contact with an architectural workshop, and I can readily believe that the practical precepts of the profession were learned by Raphael, Peruzzi, and even by Michelangelo from observing what went on in Bramante's Vatican Studio, whether or not these men were given any architectural assignments there.[6]

One can imagine a similar pattern for those who were focusing their training upon the engineering aspects of such projects. Rising demand for engineering skills, in both civilian and military sectors of the economy, eased entry into the field. Those working as engineers included Cistercian monks, masters of stonework, painters (Giotto, Francesco di Giorgio Martini, Leonardo da Vinci), doctors (Guido da Vigevano, Giovanni Fontana), sculptors (Jacopo della Quercia), and even a notary (Taccola). Their titles on the job could vary from *operaio* (operator, overseer, or contractor), *maestro* (master), *capomaestro* (chief master, in charge of the works), *architetto* (architect), and *ingeniere, ingegnere,* or *inzignerius* (engineer), according to local custom without necessarily implying any uniform level of mastery, knowledge, or skill.[7]

In the course of the thirteenth through the sixteenth centuries, engineers in Siena gradually emerged from the ranks of quasi-anonymous masters and contractors who appear in the account books or chronicles by name only to become celebrated personalities and authors for whose services patrons competed. For instance, a record from 26 September 1259 tells us only that Friar Vernaccio was from the Cistercian Monastery of San Galgano and that the city paid him £60 for work on the seats for the choir of the Duomo (Cathedral). Another document from two months later first alludes to him only by his title, then mentions his name. Vernaccio's name appears, along with his title, six times in the Biccherna volume for the first semester of 1259, with no more information than the amount he was paid in salary and reimbursed for his expenses. Even a well known fifteenth-century author such as Taccola (1385–1453) has left a slender documentary trail in the official records, for which reason his biography remains so sketchy. If his notebooks had not survived and been so often copied, he might be largely unknown today. Giuseppe Chironi's exhaustive review of the literature and archival sources related to Taccola and Francesco di Giorgio (1439–1501) offers a telling quantitative comparison of changes that took place between the 1380s and 1430s: for Taccola there are fourteen known documents, some of which are no longer extant; for Francesco there are 191 known documents, plus another four previously unpublished documents pertaining to Francesco's father, Giorgio.[8]

A review of Taccola's career and the contents of his notebooks will allow us to test the Bowsky thesis. Taccola was born in 1382, twenty-seven years after the fall of Siena's most stable government, into a period of considerable political turmoil. New governments came and went, sometimes with disastrous consequences for the people of Siena. He was baptized Mariano Danielo on 4 February 1382. His father was Jacopo, son of Vanni. His family ties placed him firmly in the mercantile and manufacturing middle classes, somewhere between the *popolo minuto* and *grasso.* We know that he worked as a wood carver because one of the earliest records we have of his professional life, from 1408, states that Taccola received a contract for sculptural work on the choir of the Duomo. How he obtained his training in wood carving is unknown; it can only be assumed that he served an apprenticeship. His contract with the Duomo set the pattern for his future employment: his patron was an institution, not an individual. Although we do not know if Taccola was ever admitted to the notaries' guild—he was nominated for entry in 1417 and 1420—it appears that whatever training he obtained in pursuit of this career qualified him to hold some minor official posts in Siena. He wrote his engineering notebook *De ingeneis* and composed his treatise *De machinis* while holding the post of *Camerarius* (secretary) of Casa di Misericordia e Sapientia (1424–1432), which was originally founded as a hospital and gradually developed into a residence for scholars. From about 1433 until 1453, Taccola used his managerial and engineering skills when he performed the duties of *viaio* (warden of the streets, bridges, and fountains) and as *stimatore* (a post combining the skills of an assessor, building inspector, and site supervisor). What part his membership in the order of the Humiliates might have played in advancing his professional career is unknown, but Frank D. Prager and Gustina Scaglia believe it to have been one of the many ways in which Taccola met important people in Siena. We do know that he had at least met some of the most important artists of his age. Jacopo della Quercia was godfather to Taccola's daughter in 1426. The personal nature of the advice Brunelleschi gave him in 1427, to keep his ideas from the public eye, suggests the regard the Florentine architect had for him. The surviving records do not indicate if Taccola ever had a private patron. The one individual to whom he dedicated his work was a head of state, the Roman Emperor Sigismund, from whom he was seeking employment. The various posts he held and imperial titles he acquired suggest that Taccola learned how to please the city of Siena and Emperor Sigismund as patrons. He seems to have done so by being attuned to the needs of his patrons, both current and potential, and responding in a manner that demonstrated his versatility.[9]

Until Prager and Scaglia's study, Taccola had been known almost exclusively as a military engineer. Prager and Scaglia have argued that Taccola was much more versatile than that term suggests and that he should not be pigeon-holed as a military engineer alone. In their publication of his *De ingeneis* they create a classification system that they believe better shows the range of his interests. They divide his work into six major classes: civil engineering, weight lifting, water supply, mills, military engineering, and architectural drawings. The first four classes are Taccola's own; to create the fifth, military engineering, they condense his six classes of military devices into just one; and the sixth, architectural drawings, they have created. Because their editions remain the starting point for any work on Taccola, I have adopted all their renderings of Taccola's technical vocabulary into English. However, it is worth remembering that Taccola himself thought it important to emphasize the variety of his military knowledge to potential patrons and made distinctions between many more classes of knowledge and devices than his modern editors.[10]

Related to his architectural drawings, but not assigned a distinct class, are the numerous decorative emblems with which Taccola ornamented his text. Taccola's designs of numerous animals included an elephant, bat, monkey, giraffe, camel, donkey, fox, weasel, fish, several birds in flight, as well as some imaginary animals like the gryphon and dragon and a personification of the wind. Such images, real and fanciful, are exactly the sorts of designs that might be found within the repertoire of a sculptor talented enough to work on the choir stalls of the cathedral. The decorative animals, which usually appear at the margin of one of his explanatory texts, are much more animated in their representation than the functional animals that appear in the main drawings, such as horses and oxen powering various mechanical devices.

Under the rubric of "Civil Engineering," which Taccola called *De pontibus et aliis editfitiis super aquas,* fall box caissons, cofferdams, pile-setting and pile-pulling devices; diving equipment; flotation equipment; harbors, harbor works, and bridge works; dredging equipment, ship-loading equipment; irrigation, swamp drainage, and regulating gates; salvage of submersed articles; and surveying and measuring. Taccola devotes at least seventy-five pages of his various manuscripts to civil engineering devices. All of the devices in this class make perfect sense when seen in the context of a city that desperately wanted its own port and that was constantly importing building materials, from its own territory and from abroad. One can infer from the drawings of salvage and lifting devices, as well as aids to diving, that Taccola thought recovering building materials and other valuable objects that had dropped into the water while loading and unloading them was a problem worthy of a solution. Considering how dependent Siena was on its roads and bridges, notes on either make up a surprisingly small proportion of the pages. Perhaps this is indicative of Taccola's goal to focus on new technology, and perhaps he felt that normal practice in road making and bridge building was adequately meeting the city's needs.

The class of "Weight Lifting," which Taccola dubbed *De ponderibus altius levandis,* included thirty-one sheets devoted to capstans and winches and twenty-seven sheets to worm drives and worm gears. Eighteen sheets relate to winches, including a winch, block, and tackle; a winch driven by weight motor; a winch driving a lever mechanism; and a winch-driven wagon on a ramp. To the other devices he devoted an average of about half a dozen sheets each. He drew bearings, camshafts, crank shafts, handles, levers, rope guides and pulleys, screws and nuts, cogwheels, gear clutches, ratchets and pawls, sprockets, spur gears, and a device intended to provide perpetual motion. In this collection of mechanical devices we can see the emerging "toolkit" of the mechanical engineer as well as the core of what would become the "textbook" solutions to the transmission of power that Eugene Ferguson discussed in his 1977 *Science* article.[11]

Of utmost interest to this study are the drawings in the class, "Water Supply," which he called, *De aquis actingendis.* The 104 pages naturally include many aqueducts, twenty-five in fact. But his frustration with the prevailing practice of depending upon gravity to deliver water can be seen in the alternatives his drawings suggest, including an Archimedean screw; seven bellows with various types of actuation; two chain pumps; twenty-four sheets of piston pumps; thirteen of siphons; twenty of water buckets on ropes or chains or wheels; and a dozen water scoops. No doubt Taccola's interest in finding better ways of raising water mirrored that of his potential patrons, be they city officials, mine owners, or individuals wanting a better fountain for their villa's garden.

The technologies employed in the class of "Mills," *De molendinis,* derives from devices that fall into other classes, especially "Weight Lifting" and "Civil Engineering." Nonetheless, Taccola still devotes more than sixty-seven pages to this group that includes such complex devices as grist mills, sawmills, and water-lifting devices. More than seven sheets represent devices driven by animal or man power and another sheet includes two tread wheels. Horizontal-shaft water wheels (and wind mills) occupy twenty-six sheets, and vertical drive shafts powered by water or wind fill another nineteen sheets. Fifteen sheets are devoted to waterpower systems, including some with reservoirs and pumps; some with tide controls; and some with floating mills. Tide mills have been in use since at least the tenth century when the Arab geographer al-Muqaddasi described one in Basra, in what is now Iraq. Al-Muqaddasi wrote, "The ebb tide is also useful for operating the mills because they are at the mouths of rivers, and when the water comes out it turns them." The earliest European record of tide mills is recorded in the twelfth or thirteenth century, significantly not on the Mediterranean, but on the Atlantic coast, with its more pronounced tides. The utility of tide mills in the Mediterranean, with its relatively minor variations of tides, is a subject of debate among historians in the field, with arguments turning on conjectural economic feasibility of utilizing small tidal movements, not documentary or archaeological evidence. The possibility that tide mills could not harness the Mediterranean's limited power further suggests a source external to the Mediterranean basin for this idea. Taccola's consideration of such applications of water power reflect not only Siena's constant struggle to ensure a steady grain supply but also the widespread trend across Europe to substitute water and wind power for human and animal muscle power in various industrial enterprises and land reclamation projects. Most of the mills around Siena were owned by the city, religious orders, or the Hospital, clearly placing mill design as another skill for which there could be potential sources of institutional or civic patronage.[12]

Prager and Scaglia's compression of Taccola's six classes of Military Engineering into one class generates at least 155 sheets illustrating this super class. The draw bridges, pontoon bridges, moats, fortress walls, and gates that occupy twenty-six sheets could arguably be called "dual-use" technology because they all have uses at times of peace as well as war. A floating bridge could ease the movement of goods after a fixed bridge washed out, something which frequently happened during spring freshets in Tuscany. The enforcement of curfews and the collection of tariffs required walls and gates, and sometimes moats, by which a city could close itself at night. Nonetheless, Taccola's own classes show that he conceived the primary use as military. Among the exclusively military technologies are assault ladders, siege towers, and battering rams; bows and crossbows; guns and cannon; gunpowder and incendiaries; stratagems; trebuchets; mines (tunnels); and three dozen devices for marine warfare. The most likely patrons for large-scale military devices would again be institutions.

At first glance "Architectural Drawings" comprise the smallest category, unexpectedly so in light of all the building programs in Siena, containing only twenty-nine sheets. However, this is perhaps the result of Prager and Scaglia's constructing overlapping categories and assigning drawings to a single category only. If one includes all the illustrations of construction technology that fall into "Weight Lifting" and "Civil Engineering," and acknowledges the architectural dimensions of "Water Supply" and "Mills," then the range and numbers of his architectural ideas are better reflected. Six sheets document the relatively recent introduction of the chimney to Italian building practice. Nine address roofs. The others contain columns, stairs, tools, battlements, fountains, door hinges, and steam blowers, which Prager and Scaglia suggest were probably intended to be installed in chimneys.[13]

Regardless of how one categorizes the individual drawings, they have in common one over-riding quality: the vast majority are devices, inventions, and practices useful to a city or a major institution within a city, not an individual. We do not find trick fountains for private gardens; we find functional public fountains. Most mills address one of the Sienese government's most pressing and intractable concerns: the maintenance of the bread supply. The military drawings show not sidearms for the defense of individuals or households, but siege engines and defenses for cities, or at the very least castles.

We know from his dedications that Taccola was trying to impress the Emperor Sigismund with the utility of his designs to the Empire. The content of Taccola's notebooks is fully consistent with the ambition he expresses in his dedication to join the Emperor's household. The evidence for Taccola's impact on the water supply system is circumstantial; all we know for certain is that he held the post of *viaio.* Except for knowing that he held that post, we do not have any direct evidence for Taccola's supervision of construction or engineering projects. What we do know is that Taccola's drawings suggest numerous solutions to Siena's chronic lack of water supply. These solutions incorporate elements drawn from the vast toolkit available to Sienese engineers, as well as the hope that by solving these problems Taccola would gain recognition from his city and perhaps a job from foreign rulers. We also know that Taccola's drawings had major influence over later engineers, particularly Francesco who definitely did apply his engineering skills to the maintenance and expansion of Siena's water supply system.

Whereas the current state of our knowledge confirms Prager and Scaglia's assessment of Taccola as an author who recorded both what he observed of existing technological practice as well as speculative recombinations of components and subsystems that he had seen or heard of, Francesco di Giorgio Martini was an author and practitioner. Taccola seems to have visited numerous construction sites and conversed with the engineers in charge, such as Brunelleschi. He likely administered some construction projects in his role of *stimatore* (surveyor) for the city government. In addition, he could have played official host for visiting experts as *camerarius* (secretary) for the *Casa di Misericordia e Sapientia* (one of the residences for university scholars), but he seems to have gotten no more hands-on experience in engineering or building than in his work as a sculptor for the cathedral in Siena. By contrast, Francesco had first-hand experience as a painter, as a military engineer on the battlefield, as a master in charge of a workshop, and as an *operaio* for Siena's water supply system. Bertrand Gille places Francesco "at the turning point" when all the elements that will constitute an engineer come together. Scaglia argues that Francesco distinguished himself both among the humanists of the *Studium* (the state-supported university) and the artists-craftsmen of the Sienese workshops. This combined fluency in craft and language is one of the distinguishing characteristics of the new professional "*ingegnere*" which in the fifteenth century came to mean "a man of unusual skills, a wizard who conceived new things."[14]

Francesco di Giorgio was baptized on 23 September 1439. Although there is no documentary evidence, many scholars have assumed that Francesco was a student of the Sienese painter, sculptor, and military architect, Lorenzo di Pietro, called Vecchietta. However, this assumption is based solely on stylistic evidence. Most of what we know about Francesco comes from Allen S. Weller's 1943 biography, which has stood the test of time quite well. However, Giuseppe Chironi's discovery of Francesco's father's tax declarations for 1453, 1465, 1468, and 1469 in the Sienese archives correct the confusion between the identities of Francesco's father and grandfather and the long-held mis-impression that Francesco's father was a poultry dealer. In fact, Francesco's grandfather had that humble occupation; his father was a minor functionary in the city government—giving Francesco a background that makes much more sense for someone who spent most of his career serving one government or another himself. Furthermore, whereas Chironi's research reveals no further documentary support for the widely held notion of his apprenticeship to Vecchietta, the tax records indicate that Francesco's home was next door to the workshop of the painter Francesco di Bartolomeo Alfei da Montalcino. The next document to record an event in Francesco's life dates from the artist's twenty-sixth year and is for a statue of St. John. The document, which records a payment of £12 for his work, refers to him as a *dipintore* (painter), a skill he could have acquired in the workshop of either master. He married in 1467, was almost immediately widowed, and remarried in 1469.[15]

The first record of his work in the field of engineering, as *operaio* on the *bottini* (the underground aqueducts) in Siena, also dates from April of that same year. The General Council of Siena commissioned Francesco, along with Paolo d'Andrea, to increase the water supply by a third over the next three years. His work required not just improving the fountains' function, but also, in 1470 and 1471, included creating special fountain displays for a public festival. The large size of the payments he received, £3200 in 1469, and another £3200 in 1472, suggests a wide scope of work and a high degree of responsibility. While employed on the aqueducts he also received commissions for some paintings, one of which may have been a map or boundary survey. He finished his first period of work on the bottini in 1472 and the records are silent until 1475 when he dissolved his partnership with the Sienese artist Neroccio. Weller speculates that Federigo di Montefeltro, Duke of Urbino, began his patronage of Francesco by then, and that this new opportunity in Urbino precipitated the end of Francesco's Sienese partnership. As his talents and fame increased, he spent more time traveling between various patrons including Alfonso, the Duke of Calabria; Giangaleazzo Sforza, the Duke of Milan; and Lorenzo the Magnificent in Florence. Francesco's work for

Federigo di Montefeltro continued until 1485, when the Sienese government requested his return. His duties in the service of the Duke of Urbino included work on the Ducal Palace, the Duomo, various fortifications, the famous carved and painted *sedile* (bench) illustrating various machines (originally located on the exterior of the Ducal Palace), as well as missions of diplomacy and espionage.[16]

Not only do we have more documentation on Francesco's career than on Taccola's, but we can also identify a large body of surviving paintings, sculptures, structures, and drawings. Among Francesco's earliest known works are miniature illustrations of the deeds of Herakles done for Alessandro Sermonetta, a teacher at the Studium in Siena, in 1463. According to Scaglia, these drawings demonstrate his familiarity with the classical tradition. Among his other early works are carved statues in wood done in 1464. He also cast sculptures, reliefs, and medallions in bronze. In addition to the buildings and structures he designed for the Duke of Urbino, Francesco also built or modified bridges, fortifications in Puglia, part of the Church of Santa Maria della Scala in Siena, and probably the church of the Franciscan monastery dell'Osservanza, just outside Siena. At his death in 1501, Francesco left a legacy of art and buildings testify to his range of interests and abilities.[17]

We can gain the best understanding of Francesco's contribution to Sienese engineering from his drawings. Scaglia's assessment of Francesco's machine drawings is that they "have little of Taccola's originality for mills, pumps, and military devices." However, "he increased the quantity of examples in each category and the combinations of existing components." Francesco demonstrates his departure from Taccola in his pumps and devices for hauling and lifting large objects.[18]

Scaglia asserts that Francesco's career as an engineer began with his work on the bottini in Siena. His most recent biographer, F. P. Fiore, argues that it was the engineering challenges posed by Siena's water supply system that brought Francesco to study Taccola's notebooks in the early 1470s. It is not clear how Francesco obtained access to the notebooks. One possibility is that they remained at the Studium in Siena, where Taccola had held the post of secretary, and that Francesco gained access to them through patrons such as Sermonetta, who taught there. In any case, we know that Francesco had access for the following reasons. First, Francesco made annotations on the last pages of Taccola's notebook *De ingeneis,* as Michelini recognized in 1962. Second, machines based on *De ingeneis* appear in the autograph copy of Francesco's *Opusculum de architectura* (1470–1475). Also, copies of Taccola's machines and translations of Taccola's Latin text appear in Francesco's *Codicetto,* his pocket-sized sketchbook of machines from 1470–1490.[19]

Although we cannot directly link Taccola to the design or construction of Siena's water supply system, the utility of his drawings is apparent in Francesco's use of them, starting within a year or two of when Siena hired Francesco to work on the water supply system in 1469. Among the pages of Taccola's notebooks on which we find Francesco's annotations is folio 130 of *De ingeneis* I-II. On this page Taccola illustrates floating mills, an aqueduct carrying water to what appears to be a grist mill, a windmill powering a noria connected to another grist mill, a device for transporting a large architectural column, and another device that appears to be designed for lifting a sunken column from beneath the surface of water. Francesco's notes translate Taccola's description of column-transporting and column-lifting devices. Historian Paolo Galluzzi has asserted "no document better expresses the continuity of the Sienese tradition of machine studies. They offer us a snapshot, so to speak, of the actual moment when that heritage was passed on from Taccola to Francesco di Giorgio." Although Galluzzi goes on to speculate that they "may even indicate a brief apprenticeship of Francesco with the elderly Sienese engineer, which in chronological terms is not impossible," the annotations alone demonstrate the transmission of Taccola's ideas to Francesco and perhaps their application to the Sienese water supply system.[20]

Francesco's copies of Taccola's drawings are important because they are disseminated through at least three paths of transmission. First, as part of the ongoing manuscript tradition the reader will recall that Francesco's manuscripts are the most copied manuscripts of any artist of the fifteenth and sixteenth centuries. Francesco's audience included no less an engineer than Leonardo da Vinci. As Ladislao Reti observed in 1964, notes and drawings in Leonardo's own hand appear in the margins of Francesco's *Trattato di Architectura.* Second, Reti goes on to point out that it is more likely that the vocabulary of machines from the Sienese school made its way into the Theater-of-Machines books from Francesco's and not Leonardo's drawings. Third is the avenue that Chironi is pioneering, which is to determine the extent that Vannoccio Biringuccio's *Pirotechnia* has depended on Francesco's ideas. Of the three, Chironi's work promises the most new information regarding the social and economic context in which Sienese engineers worked.[21]

Chironi's research has increased our knowledge of the connections between Francesco di Giorgio and other prominent Sienese citizens, including Paolo di Vannoccio Biringuccio, the father of Vannoccio Biringuccio (1480–1537), the author of the *Pirotechnia.* Chironi has published documents from the Archivio di Stato of Siena detailing Francesco's and Paolo's service in 1487 on the governmental committee in charge of roads, bridges, and fountains, called the *Viari.* In subsequent documents Chironi shows that

Paolo and Francesco obtained approval from the Sienese government to form a partnership in 1488 along with one of Siena's most powerful citizens, Pandolfo Petrucci (the future tyrant), and Paolo Salvetti, to build "mills and various other structures" along watercourses in Siena. Chironi's findings are important because they give more clues as to the possible details surrounding the transmission of Taccola's and Francesco's drawings into the printed machine books. One of these ideas—the construction of a mine to blow up an enemy fortification—occurs as an idea in Taccola's *De machinis,* is put into practice by Francesco and is published by Biringuccio, who credits Francesco directly in Book X of his *Pirotechnia.* Chironi's findings are also important because they show that Francesco was sitting on committees and making business deals with members of Siena's ruling elites.[22]

The rising status of engineers resembles that of painters and sculptors during the same period. Art historian Arnold Hauser characterized the changing status of the artist in the following manner:

> The increased demands for works of art in the Renaissance leads to the ascent of the artist from the level of the petty bourgeois artisan to that of the free intellectual worker, a class which had previously never had any roots but which now began to develop into an economically secure and socially consolidated, even though by no means uniform group.[23]

Engineers were like architects, with whose skills and tasks they overlapped; unlike artists, they neither had a guild, university examination, or other ready means of defining the boundaries of their occupation in the thirteenth and fourteenth centuries. For example, Taccola worked as a sculptor and applied for admission to the notaries' guild, but there is no record of his having succeeded in becoming a member of that or any other guild. Even the otherwise well documented career of Francesco di Giorgio has serious gaps in the first two decades of his life. The earliest record of Francesco's professional activities refers to him as "painter," but we have no record of his apprenticeship or admission to a guild. Perhaps the weakness of the guilds in Siena, compared to Florence where they played a much larger role in the government, made it easier for people with related skills to expand the scope of their occupation to include engineering. The weakness of the guilds could also have made it easier for foreign-born engineers to work in Siena.[24]

One of the defining steps toward greater recognition of the occupation of engineering and the acquisition of engineering skills by a particular practitioner was the preparation of collections of engineering drawings in the form of bound notebooks for presentation to a potential patron. Taccola was a pioneer in this new medium and in this new role of artist-engineer as author. It was not until the following generation that authors such as Francesco di Giorgio succeeded in supporting themselves as artist-engineers full-time. Francesco not only refined the medium further through better drafting and framing devices, but he also lived out the role more fully, supporting himself almost entirely from artistic and engineering commissions as well as investing in some milling partnerships with powerful Sienese citizens. In this way Francesco di Giorgio's career resembled those of the architect builders who oversaw the construction of the great cathedrals of the thirteenth and fourteenth centuries. Richard Goldthwaite, in his studies of architecture and the construction industry in Florence, has articulated important questions about the nature and effect of various forms of patronage upon the building arts. Goldthwaite finds much more widespread participation in the design, finance, and construction of public monuments, including cathedrals, in Italy than north of the Alps, where the nobility frequently excluded non-nobles from cathedral chapters. One can only speculate upon the impact of having non-nobles who made their fortunes in commerce and other businesses working closely with the master builders and engineers on a cathedral's business.[25]

The experience engineers gained in Siena and other cities formed the basis of a technological legacy they bequeathed to the rest of Europe. In addition to the tradition of engineering notebooks that Ladislao Reti has traced back to Taccola, we need look no further than the cathedrals, city halls, public fountains, clock towers, rings of walls, and private palaces that dot the surviving medieval urban landscape to see the impact of urban patronage upon the advance in engineering skills. The city-state in its role as patron was the late-medieval precursor to the more organized efforts of the French government to establish engineering schools such as the Académie Royale d'Architecture in 1671, the Corps du Génie in 1676, Mézières in 1749, and the École des Ponts et Chausées in 1775. Frederick B. Artz writes that the Académie Royale d'Architecture was "as much a school of engineering as of the fine arts" and goes on to call it the "first higher technical school in France." The underlying structures by which the water supply system and other public works were built constitute the lasting legacy of the aqueducts and fountain-building process in Siena. The keystone of this system was the creation of a new social role, that of the full-time urban engineer. The role of the medieval engineer and his relationship with urban patrons represented a new structure. The rich documentation of the water supply in Siena presents an opportunity to describe a fundamental shift in the relationship of technology and society and thus to fulfill a part of White's agenda by contributing to our knowledge of engineering in Siena.[26]

Despite rich documentation for the thirteenth and later centuries, when medieval cities were reaching their peak populations, evidence for the means and structures by which large public works were built before the re-urbanization of Italy is scant. The fourth through ninth centuries have left an especially thin record. We are forced to infer from the better-documented later centuries what happened in earlier periods. For instance, it would appear that before they were very large, cities initially borrowed skilled members of craft organizations from the large country estates, mines, monasteries, and cathedral workshops. Borrowing experts on an ad hoc basis was adequate to meet the sporadic needs of small medieval cities. For example, the government of thirteenth-century Siena often consulted experts from both the nearby Cistercian monastery of San Galgano as well as from the cathedral works of the city itself. A good example of the close cooperation between San Galgano and the city was the transitional period between 1258 and 1285, when the Master of the Cathedral Works was always a monk from San Galgano. As cities grew in size and complexity and as they became more numerous, their engineering demands became more frequent and specific to their urban form and scale. A new social order of expertise ultimately evolved to meet the special needs of cities and their rulers. Engineers became essential if rulers and cities were to preserve their autonomy.[27]

The ongoing municipal investment in public works projects in Siena had the effect of expanding its pool of ecclesiastic and military engineers whose task it was to design and direct such projects. This group of engineers provided a body from whom could be nominated members of an ad hoc advisory committee, such as the panel of experts Siena assembled to review the plan for its cathedral in 1321. The panel consisted of five consulting architects, including three "foreigners" from Florence, to evaluate the structural stability of the expansion then in progress. The composition of these ad hoc committees attested to the ready availability of masons, architects, and engineers with sufficient expertise to inspect a project as complex as a cathedral's structural integrity. When Milan was searching for a new chief architect to replace its French engineer, Nicolas de Bonaventure, at the end of the same century, it cast its net as far afield as Ulm, Gmünd, and Cologne. Sienese engineers such as Jacopo della Quercia and Francesco di Giorgio routinely traveled to other cities on the peninsula to supervise major projects.[28]

Siena's greatest period of sustained economic and physical expansion took place in the thirteenth and fourteenth centuries. Most of the monumental structures that have come to define Siena's medieval identity date from these two centuries. They include the Piazza del Campo; the Palazzo Pubblico; the Torre della Mangia; the first and second enlargements of the cathedral; the two large churches of the preaching friars, San Domenico and San Francesco; the monastery of San Galgano; much of the Ospedale and its dependent farms, granaries, and mills; the port at Talamone; the fountains of Branda, Nuova, Ovile, Peschaia, and Follonica; and numerous roads, bridges, dams, aqueducts, and drains that knit the infrastructure together. When we add the construction or expansion of numerous private palaces, we can begin to see the amount of sustained patronage that just one city offered its designers and builders. Despite the June 1348 outbreak of the Black Death that caused the work on the second cathedral expansion to stop abruptly, work on public projects soon resumed. The Chapel in the Campo, built by survivors of the Black Death (1376), the Fonte Gaia (1409–1419), and the Loggia della Mercanza (1430s) all date from the century following the plague.[29]

In his study of the effects of the plague in Siena Bowsky found a pattern of unexpectedly resilient civil institutions recovering quickly from the staggering death toll of the plague of 1348. One thing he did not find was any legislation that seemed to have been immediately enacted as a result of the plague of 1348; the plague is reflected only in later Sienese statutes and those of other cities, like Pistoia. The lack of surviving ordinances in Siena could be simply a result of the confusion during the plague or the revolution seven years later, when so many records were burned. It is possible that official measures in Siena resembled those in Florence. According to Giovanni Boccaccio's account, filth was removed, sick people were prohibited entry to the city, and many ordinances were issued concerning the maintenance of good health. Insofar as the plague only briefly interrupted construction in Siena, it appears to have had little permanent impact on the development of engineering in the city.[30]

The burgeoning body of knowledge that the engineers collectively husbanded coincided with the development of a complex of technical innovations by which engineering, architectural, and scientific ideas could be recorded with greater accuracy and disseminated more quickly. According to Ferguson, improved methods of representation such as pictorial perspective and "exploded" diagrams of machine parts were more important to the transmission of technical ideas than the ability to mass-produce text. The emphasis on images diminished the role of verbal communication and lowered the barriers that vernacular languages might have once presented.[31]

The formation of a larger, international community of practitioners began a significant phase in the history of engineering. Word of large projects spread quickly and drew onlookers. Surviving sketchbooks dating back to those of Villard de Honnecourt (active 1225–1250) and numerous other builders attest to the curiosity practitioners had in each other's work. Ackerman believes that such sketch books functioned as "textbooks for architectural training" by extending the knowledge of students to monuments they could not visit in person. Brunelleschi's counsel to Taccola, to keep his ideas from the public eye, highlights some of the risks onlookers posed to what would now be called an engineer's intellectual property. This body of knowledge, enhanced by new techniques of representation and communication, accelerated the pace of technological innovation and the application of new solutions to old problems. The creation of an engineering community was perhaps more important than any of the particular technical solutions discovered at the time.[32]

The new engineering community was not constrained by political or linguistic boundaries. Thus, individual engineers who were in direct contact with each other in central Italy were integrated into a larger network of experts, whose knowledge, techniques, and experience stretched across Europe and beyond. One copy of Taccola's *De machinis* even found its way to the library of the Grand Sultan, Mehmed II (1432–1481), in Istanbul. European engineers formed a network composed of far-flung practitioners, many of whom would never meet in person, but who, through improved communication, would begin to function as a larger body. Ferguson writes that in this way "the circle of technologists engaged by a particular problem or stimulated by a particular idea was thus indefinitely enlarged." With this greater potential to solve problems came the power to disrupt the status quo; thus, engineers' talents were often met with some misgivings.[33] White, although writing of a later period, suggests that contemporary perceptions of their effect upon society were not completely positive:

> For nearly a century thereafter [i.e. the sixteenth century] Italian engineers scattered over Europe, from Madrid to Moscow and back to Britain, monopolizing the best jobs, erecting wonderful new machines, building palaces and fortifications, and helping to bankrupt every government which hired them. To tax-paying natives they were a plague of locusts, but rulers in the sixteenth century considered them indispensable.[34]

Nonetheless, their utility to urban rulers guaranteed the survival of engineers.

Communication among practitioners could take many forms. It could be local and in person, such as the meeting in Siena in which Brunelleschi advised Taccola to be secretive with his ideas. It could be through the transmission of diagrams or formulas, as we see recorded in Francesco's notes on pages of Taccola's notebooks, and in those of other engineers that have copied or referred to Taccola's diagrams. Indirect communication, such as that afforded by rumor, could be a stimulus to innovation, as well. Alfred Kroeber's stimulus diffusion model, addressed in Chapter 5, could be equally applied to modes of intra-European diffusion. The mobility of engineers guaranteed that overtly competitive efforts, which were not intended to be cooperative or collaborative, such as the race between cities to build the largest cathedral or the best fortifications, could have the unintended result of improving the level of collective expertise. Through their long-distance collaboration, more minds could be directed at persistent problems such as those arising from construction, fortification, water supply, or the application of power to new kinds of mills. Public works projects across Europe could function concurrently as laboratories in which these ideas were tested. It is a testament to the rapidity with which new technical ideas spread through Europe that the first instance of an innovation, such as clocks, is so hard to pin down.[35]

Prager and Scaglia's study of Taccola not only represents a huge leap forward in understanding Taccola and his context, but it has also stood the test of time; little new information concerning him has emerged in the three decades since its publication. They recognize the effect Siena had on Taccola, arguing that Siena's hilltop location, and the impediments its very real shortage of water placed in the way of applying water power to industrial processes, helped shape his agenda. The constant threat marauding mercenaries posed to Siena's physical security and political and financial stability stimulated his quest for better military strategies, defensive measures, and engines of war. And certainly the unceasing efforts of the city to improve its appearance provided plenty of opportunities for sculptors, painters, and master builders like Taccola, Francesco di Giorgio, and Jacopo della Quercia to practice their arts. Siena's many commercial, intellectual, and military contacts with the world beyond Europe permitted a steady stream of new ideas and technologies to reach even its inland location. These conditions are necessary but not sufficient to explain why artist-engineers like Taccola and Francesco di Giorgio are from Siena and not Gubbio, Arezzo, Volterra, Orvieto, or other small hill towns facing similar challenges to their industrial capacity, military strength, or water supply.

Bowsky's research sheds new light on White's question: Why Siena? Bowsky argues that through their long tenure, clearly defined goals, and assiduous attention to detail, the Nine set a pattern of patronage and rewards that would endure long after their reign had ended. In his article, "The Flavor of Early-Renaissance Technology," White stressed the important role of patrons in fostering the growth of the engineering community. The Nine, according to Bowsky, created a set of financial, social, and political rewards for public servants, including artists and engineers, who worked for the common good, analogous in ways to the patronage of dukes and princes who ruled cities. White suggests that the innovative quality of Renaissance engineering "reflected the new needs of a rapidly expanding commercial society and of monarchs eager to increase not only their armed might, but also the prosperity of their realms." Citing some of Leonardo's patrons as examples, White continues, "Burghers and tyrants did indeed value engineers." The competition between Siena, Urbino, and other cities for Francesco's talents bears White out on this point. Weller writes, "In the years of his greatest fame, Francesco was the object of considerable haggling between the various governments that desired his services."[36]

By virtue of their scale, medieval cities were quantitatively and qualitatively different from most other medieval patrons except perhaps the very largest monastic enterprises. They required a wider range and a greater quantity of engineering services than the dioceses or manors that bishops or rural lords had governed in previous centuries. The scale, continuity, and access to resources that characterized cities fostered an environment well-suited to training and recruiting engineers. In turn, a city could benefit by accumulated expertise and institutional memory among the engineers it employed long-term. The continual need of cities to repair, replace, defend, and expand their infrastructure offered engineers employment for years at a time. For instance, building the public structures surrounding Siena's Campo complex spanned the decades from the 1290s until the 1350s. By the time Jacopo della Quercia completed the Fonte Gaia in 1419, the city had been developing its most important zone, the Campo, for at least 130 years. Such projects allowed several generations of engineers to be trained in the course of their completion. Even without a particular major project, the ongoing needs of maintenance and even assessing taxes could provide someone with engineering skills steady employment. The reader will recall that Taccola spent the last twenty years of his life as a city official concerned with buildings and roads. Not only did the city need artist-engineers but, as Bowsky points out, the city rewarded them generously encouraging them to remain in Siena instead of having to seek work elsewhere.[37]

An examination of biographies of artists and engineers and data contained in public records suggests that skills and know-how were frequently transmitted via apprenticeship. Insofar as membership in artists' guilds was largely hereditary and *botteghe* (workshops and studios) were family businesses, skills were often passed within families and households. Because the qualifications of an engineer or architect were still being negotiated, however, and no guild for architects or engineers existed, paths to entry varied. Engineers could come from either workshop traditions as did Brunelleschi, Francesco di Giorgio, or Leonardo, or from traditions of learning based outside the workshops, like Taccola, who attempted without success to enter the notaries' guild. In a time of flux, extended projects such as the Campo and the cathedral lent stability to the process of apprenticeship and would create an ideal environment for transmitting and augmenting such skills across disciplines and over several generations, by bringing together allied professions "under one roof." In addition to those who directly benefited from being employed on a project, knowledge spread to those citizens who participated in advisory panels—many of whom were chosen because they were artists or because they possessed some technical knowledge. As a patron, Siena used fiscal and political resources comparable to those of many tyrants to advance the interests of the burghers who wielded so much influence in its various governments. The larger pattern of re-urbanization resulted in cities' and urban patrons' gradual supplanting of rural monasteries and rural lords as the leading patrons of architecture, arts, and engineering.[38]

The engineering challenges posed by large public works projects did not cease with the completion of construction. The nature of the aqueducts, mills, bridges, clocks, and large public buildings was such as to require constant maintenance. No sooner was the first fountain in the Campo finished than residents of abutting neighborhoods wanted more water and more fountains. The complexity of maintaining the new urban infrastructure required highly skilled supervision, a task well-suited to Taccola and his contemporaries. The inherent complexity of urban systems also created a bureaucratic imperative for a city to employ the same engineers over longer terms than those of its elected officials. The post of operaio of a fountain or aqueduct was specifically exempt from the term limits imposed upon many other Sienese officials. Capo maestri of the cathedral enjoyed the same exemption. From 1259 until 1537 fewer than two dozen men held this post. In those same three centuries fewer than four dozen men served as operaio del Duomo. The city understood that the accumulation of an institutional memory and skill base was more important than the benefits of more rapid turnover in such offices.[39]

Urban governments often brought several engineers together as full-time directors of projects, as judges and entrants in competitions, or as consultants. Taccola and Francesco both supervised the administration of the water supply system in Siena. Brunelleschi, Ghiberti, and the Sienese artists Jacopo della Quercia and Francesco di Valdambrino competed for commissions such as the one to design and cast the bronze doors of the Baptistery of S. Giovanni in Florence. Competitions often forced them to reveal their latest ideas to each other. When the extension of the Duomo in Siena showed signs of structural failure soon after construction began in 1316, the city convened a group of consultants from Siena and Florence on 17 February 1321. Such conferences brought engineers from other cities into face-to-face contact in ways that seldom happened before cities began to grow in the Middle Ages.

The numerous physical requirements of cities provided a range of projects, reflected in Taccola's notebooks, that few individual patrons could offer. Some of the more famous individual patrons, such as the Medici of Florence, the Sforza of Milan, the Montefeltro of Urbino, and the popes in Rome were important patrons precisely because they were, in fact, rulers of cities. The majority of their engineering needs stemmed from this administrative role. Finally, cities possessed the financial mechanisms by which to bring huge projects to completion.[40]

Since Bertrand Gille's work in the 1960s, scholars have recognized that the roots of what we know as "Renaissance engineering" lay not in the sudden appearance of seemingly lone geniuses like Leonardo da Vinci. An examination of the water supply in Siena allows us to focus instead on the new intellectual, cultural, and social environment that developed in the Middle Ages—an environment shaped by synergies between urban patrons and dynamic economies of trade and manufacture—and an expanding group of engineers eager to meet the challenges these changes posed. The innovations of printing and pictorial perspective sped the dissemination of ideas. The impact of these synergies would not be felt immediately, but as they grew in scope and significance, the community of engineers in Europe would come to shape irreversibly an ever-growing portion of the known world.[41]

Chapter Seven

Conclusion

"It early occurred to me that since there was a real possibility that all three of these innovations that had liberated Europe from medieval bondage [the magnetic compass, gunpowder, and printing], and especially the first two, had reached the West from China, the inspiration of the Italian Renaissance might have been Peking as much as Rome or Athens; but firm data seemed hard to get and I never developed the idea." (Lynn White, Jr., "Study of Medieval Technology")

In reviewing the place of Siena in the history of water supply, and more broadly, urban engineering, three themes emerge. First, an urban demographic explosion that began around the tenth century introduced both constraints upon and opportunities for urban rulers. Second, this urban growth spawned the evolution of the water supply system in response to demographic pressure. Third, the development of engineering as a distinct, specialized occupation, was to reach its fullest flower as cities expanded and attained such a scale that new technical knowledge was required. Therefore, the role of the "Renaissance engineers" was to seek and adopt techniques that had supported large populations–in classical antiquity, in monasteries, and in foreign cities as well as to invent new means from scratch. Instead of echoing the Humanist rhetoric of a "revival of ancient learning," scholars in fields such as the history of literature, science, and technology have begun to acknowledge extra-European sources as being among the most important of all throughout the era that had once been widely heralded as the Renaissance.[1]

When I began this project, it had been my hope that a close examination of the water supply system of Siena could offer a case history demonstrating the sort of diffusion which Lynn White, Jr., had demonstrated for the ogival arch and had so long sought for other techniques. The Sienese aqueducts' physical resemblance and functional similarity to qanats, and its location in an area that had received and further developed so many other inventions from the East suggested it would be possible to consider the bottini of Siena as part of a large technological system that could have been brought to Europe with more easily traced eastern exports, such as rice, paper, gunpowder, Hindu-Arabic numerals, and silk. One clue that this task would be more difficult than anticipated was the strange local name for the system. The word *bottini* appears to bear no relationship to canal, cuniculus, or qanat, all of which attest to their common origins in antiquity through linguistic similarity. In addition, the lack of "firm data" with respect to the origins of the city of Siena, let alone its water supply system, required the construction of a diffusion model sufficiently flexible to account for pre-Roman, Roman, or Islamic origins of the technology, as well as independent invention.

A more fruitful result of this research has been the integration of "internalist" histories (those which focus on technology in isolation from social and cultural contexts) of the water supply system in particular and public works in general with biographical data of some of the better-known Sienese engineers. By examining the careers of Mariano di Jacopo detto il Taccola, Francesco di Giorgio Martini, and Vannoccio Biringuccio in light of William Bowsky's thesis that Siena, more than other Italian communes, reserved its highest rewards for those artists and public servants whose work most benefited the common good, we can suggest an answer to White's earlier query as to why Siena provided such a fertile place for engineering. By placing our knowledge of Sienese engineers into a broader social and cultural context, this study offers a new synthetic explanation of that intellectual climate. Siena's lack of an adequate water supply provided an ongoing technical challenge, its system of patronage provided material support, the workshop system afforded opportunities for training, and the city's government and culture provided material and social rewards to those who could help solve its technical challenges.

Although providing a synthesis and interpretation of existing literature, this study has not extended its scope to include dozens of other master builders, engineers, and architects who did similar work on Siena's public works and monuments because the evidence for details about their careers is so thin and that because it is not yet clear what questions the data could answer. Until the authors of some of the anonymous Sienese manuscripts can be identified, we will not have the sort of internal evidence that the literary productions of Taccola, Francesco, and Biringuccio offer us. We need only recall that the vast majority of what we know about Taccola is inferred from his notebooks, not deduced from the meager documentary trail he left behind. In the meantime, it is hoped that questions presented in this work will inform such research.

The physical evidence of the known and unknown engineers and artists may help to fill in some of the details the documents lack. One of the most appealing aspects of studying the Sienese water supply system is that it is more than 90% intact and is still in use. Even so, the process of mapping, recording, and analyzing its vast physical evidence only began in the 1990s. Likewise, Siena's cathedral and numerous churches are only in the early stages of being catalogued, let alone investigated archaeologically. The greatest promise for a deeper understanding of medieval and early-modern Siena rests in the use of the physical data to formulate better questions, ones which will allow the documentary evidence to be more fully exploited.

A complementary requirement for the continued study of Siena's archaeological remains and the built environment is the assessment of the upstream and downstream effects of water supplies, industries, and cities upon the natural environment. Conducting this research will require collecting evidence such as soil cores, pollen samples, remote sensing, and archaeological artifacts that has, for the most part, not yet begun for the medieval period in Italy, especially for its cities. Future research, along the lines of Richard Hoffmann's magisterial work on mill and dams in Northern Europe, will allow us to understand the impacts of urban water supply systems and industry upon ecosystems in the Mediterranean Basin.[2]

A discussion of sewage lies outside of the scope of this work because the volume of the new water supply carried by the *Bottino Maestro* was not great enough to require new methods for the disposal of waste. In this way Siena's system was unlike those higher-volume, pressurized modern water supply systems that necessitated new ways of removing waste water. Until the early twentieth century, when Siena finally built sewers at the same time it built a pressurized water supply system, water continued to be hauled up stairs on the backs of women and dumped out windows or drained through holes in the wall or lead pipes to the street below. Because households that abutted the new aqueduct connected their cisterns to it, they precluded the practice common in cities with new water supplies, of reusing cisterns as cesspools. Consequently, except for documents and structures already cited, my research uncovered no information about earlier systems of waste disposal.

Among the other currently unanswerable questions are the effects of the Black Death of 1348 upon the water supply system of Siena. The state of the research right now is such that we are less sure of the origins, biological causes, and effects of disease in the Middle Ages than we have ever been. There continues to be a need for local investigations of the plague in Siena like Bowsky's 1964 study. The effects of the plague in Siena are particularly hard to determine because the epidemic was so soon followed by the city's most profound political revolution, in 1355. The archives barely speak of the plague; it is more noticeable in omissions of meetings or regularly recurring events than in records of its impact. For now, the only thing we can say for certain is that it killed around half the city's inhabitants between May and August of 1348. New techniques, such as archeogenetics, and collaboration between historians and researchers in genomics promise more robust diagnoses and a greater understanding of the health of human populations. Recovery of plague DNA from four hundred year-old human dental pulp in Marseilles suggests the power of such techniques to understand human-microbial interactions and the place of our species in the natural world.[3]

The direct effects of expanding the water supply system in Siena seem to have been as much symbolic as practical. Most of the new water supply went to domestic use and public display. The impact on domestic users was one of greater convenience and lower costs, but these benefits were confined to those few wealthy households that were able to connect to the new aqueduct, perhaps one percent of the population. The effect on the city was to provide its greatest public space with the Fonte Gaia, a symbol of the city's ingenuity and unity.

It is difficult to separate the art/society and technology/society relationships in Siena because so many of the patrons and practitioners were involved in both. Sienese artists, engineers and their patrons seldom made that distinction themselves. Art and engineering were viewed as important craft activities vital to a city's health and beauty. The city took care to nourish and retain talented workers, and both art and engineering flourished under the city's wise patronage. Part of the city's ability to provide appropriate support and maintenance to the arts and engineering lay in its recruiting craftsmen to join the committees that made the city's day-to-day decisions about its physical environment. This involvement of the very people who shaped and maintained the city's infrastructure contributed to a lasting legacy that is still visible today. Cities aspiring to be livable and beautiful have much to learn from Siena about the value of enfranchising of citizens of every rank, rewarding civic behavior, the collaboration of arts and technology, and building infrastructure for the long term.

Notes

NOTES TO CHAPTER ONE

1. White, "Flavor of Early Renaissance Technology," 41–42. Bortolotti identifies at least six additions to the walls between 1100 and 1420 in *Siena,* 6, 27–29, 60.
2. Falletti-Fossati, *Costumi Senesi.* See for instance Hall's paean to Florence, in *Cities in Civilization.* Douglas, *History of Siena.* Schevill, *Siena.*
3. The best example of the "exceptional attitude," which has, in turn, set the tone for almost all subsequent studies, is Bargagli-Petrucci, *Fonti di Siena.*
4. Bowsky, *Medieval Italian Commune,*19, gives 1328 as the year of peak population.
5. For Maitani's work on the facade of the Cathedral at Orvieto see Burckhardt, *Siena,* 50. For Lando in Naples, see Hook, *Siena,* 64.
6. Herlihy, *Black Death,* 39.
7. For a somewhat credulous recounting of these mythic origins, see Douglas, *History of Siena,* 4–14.
8. Hook, *Siena,* 7, 8. Bowsky, "Siena," 278; idem, *Nine,* 85. *Enciclopedia italiana,* s.v. "Siena," 719. Schevill, *Siena,* 51–67. See Waley, *Italian City-Republics,* 1–10, 32–40. Painter, "Western Europe on the Eve of the Crusades," 11.
9. Schevill, *Siena,* 67. Waley, *City-Republics,* 207. Bowsky, "Siena," 278.
10. Bowsky, "Siena," 278–279; idem, *Nine;* idem, "Black Death," 2. Wieruszowski, "Art and the Commune," 14–33.
11. All figures are given in *Lira* (abbreviated here as £), the money of account, which steadily lost value against gold florins during the period under discussion. *Lira, soldi,* and *denari* are calculated in the same way as pre-decimal British pounds, shillings, and pence: £1= 20s.= 240d. Bargagli-Petrucci's total for the fifteenth-century fountain and aqueduct expenses is £166,387, suggesting perhaps a slightly slower rate of spending after 1400. Unfortunately, Bargagli-Petrucci only broke his figures into fifty-year intervals, so we cannot do a year-by-year comparison without consulting the manuscript account books. Bargagli-Petrucci, *Fonti di Siena,* 1:132–133.

Data are available from surviving Biccherna volumes for seventeen of the twenty-eight semesters between 1286–1300, including both semesters for 1286, 1288, 1289, and 1292. Bowsky, *Finance,* 298–299. Braunfels, *Urban Design,* 61–67.

12. Bowsky, *Nine,* 185.
13. Goldthwaite, *Building,* 385–388.
14. Cecchini, "L'Archivio dell' Opera del Duomo, 67–69.
15. Although the old building was vacated in the 1990s to form the core of a museum complex, the hospital continues in operation to this day, functioning as a major center of medical research and education. The management of the botanical garden that it once administered has been transferred to the University of Siena.
16. Bowsky, *Nine,* 277. Prager and Scaglia, *Mariano Taccola,* 11.
17. Goldthwaite, *Building,* 1, 2, 8, 19, 174.
18. Goldthwaite, *Building,* 422–425.
19. Garnsey, *Cities, Peasants and Food,* 74. Statuto 5 (1297), and Statuto 20 (1309–1310). The latter is reproduced in Braunfels, *Mittelalterliche Stadtbaukunst,* 250. Galluzzi, *Prima di Leonardo;* idem, *Renaissance Engineers.*
20. Renouard, "Peste Noire," 111; Bowsky, "Black Death," 17, 19, 32–34. Caferro, *Mercenary Companies,* xiii-xiv. Hook, *Siena,* 52; Bowsky, "Siena," 280. Douglas, *Siena,* 151.
21. Schevill, *Siena,* 219–221, 227. Bortolotti, *Siena,* 22. Bowsky, "Siena," 281–282. Hook, *Siena,* 172–173.
22. Bowsky, "Black Death," 19.
23. "I Dodici Governatori della Repubblica did Siena deliberano che il Duomo Nuovo perchè minaccia rovina, sia disfatto." Archivio dell'Opera del Duomo di Siena. Libro di Documenti Artistici N. 6, in Milanesi, *Documenti,* 254–255. Burckhardt, *Siena,* 66.
24. In fact, the city the Nine built might be even more visible today than it was in 1900 thanks to efforts during the past century to "medievalize" the historic center. The elimination of twentieth-century intrusions such as overhead wiring and motorized traffic have reduced obstructions, while medieval reconstructions of numerous sites further emphasize the achievements of the Nine. For a deeper discussion of this trend, see Van der Ploeg, *Art, Architecture, and Liturgy,* 11–12. Van der Ploeg contends that the "history of Siena's medievalization has yet to be written."
25. Bargagli-Petrucci, *Fonti di Siena* 1:214; 2:190–191.
26. Luchini and Ligabue, *Siena dei bisnonni,* 250–251. We could speculate that much of Siena's late nineteenth-century squalor resulted from its inability to reap any of what the economist Alexander Gerschenkron called "the advantages of backwardness." Perhaps Siena was a latecomer to the water-supply technology of the industrialized world precisely because it had once been so

advanced. Gerschenkron, *Economic Backwardness.* Thanks to Prof. Reed G. Geiger for providing the reference.

27. Zdekauer, *Constituto.* Bowsky, "Sienese Archive," 69, 71. Bowsky notes that the vast majority of these volumes cover the thirteenth to early fifteenth century. Milanesi, *Documenti.*
28. Bowsky notes his debt to Wieruszowski, "Art and the Commune," and Larner, *Culture and Society in Italy,* in shaping his discussion of art in Siena. Bowsky, *Nine,* 260, 284, 298. Prager and Scaglia, *Mariano Taccola,* 17, 154. Weller, *Francesco di Giorgio,* 19, 22, 34. Chironi, "Politici e ingegneri, passim. Fiore,"Francesco di Giorgio," 754. Tucci, "Biringuccio," 625.
29. Bargagli-Petrucci, *Fonti di Siena.*
30. See for example, Riccetti, *Città costruita* and Feruglio, "Approvvigionamento idrico nell'antica Perugia," 217.

NOTES TO CHAPTER TWO

1. Sestan uses the phrase "figlia della strada" in "Siena avanti Montaperti," 30–33. Bezzini affirms Sestan's ideas about the crucial role of the road networks in *Formazione e sviluppo,* 1–18.
2. William Bowsky, "Siena," gives 1348. See also *CS,* 130. The equal heights of the two towers is an observation I owe to Sienese artist Bernardo Giorgi who argued this point in his Baccalaureate thesis. Personal communication, November 1993.
3. This elegant expression of scale is from Waley, *Siena and the Sienese,* 1.
4. Lawrence, *Etruscan Places,* provides a striking description of the misery malaria continued to spread along the Tuscan coast well into the twentieth century.
5. Owens, *The City in the Greek and Roman World,* 13–14.
6. For more on the urban churches of the Franciscans and Dominicans, see Braunfels, *Monasteries,* 125–152. A complete architectural survey of churches in Siena, Riedl and Seidel, *Kirchen von Siena,* was begun in 1985.
7. Italy, Servizio Geologico, *Carta Geologica d'Italia,* sheet 2 and text 2. Tufa should not be confused with tufo, or tuff, words denoting a much harder rock of volcanic origin. Tuff is also called tufa in English, and in most of Italy *tufo* means tuff, but in Tuscany *tufo* refers to the local sedimentary rock commonly called tufa in English. The geologic terms would be confusing enough had not the Etruscans made drainage systems in central Italy by cutting tunnels through the volcanic tuff. Ward-Perkins, "Etruscan Engineering." The frequent cave-ins that made digging the aqueducts difficult and dangerous were often due to the weakness of this material. This aggregate's instability could have been one of the reasons for the settling of the foundations of the cathedral during its construction.
8. Hughes, "Land and Sea," 90.

9. Schevill, *Siena,* 4. Bortolotti, *Siena,* 3. Bortolotti's profusely illustrated history offers the best overview of Siena's physical development.
10. Bortolotti, *Siena,* 5. Gams, *Series episcoporum ecclesiae catholicae,* 752, provides the following sequence: c. 309 Lucifer(i)us; 465 Eusebius; c. 635 Maurus; 670 Gunteranus (Gualteranus); 678 Vitalianus; 700 Magnus. Similar gaps appear in Gams's series of bishops in the neighboring dioceses of Arezzo, Chiusi, Lucca, Florence, Grosseto, and Volterra.
11. Cristofani, *Siena,* 18, 185–212. A close examination of the catalog of artifacts accompanying a map purporting to represent a rich lode of archaeological finds in Siena shows how thin and conjectural the evidence is. For instance, figure XXVIII, "Carta dei rinvenimenti in Siena città" (195), shows a group of objects of uncertain provenance recovered over a period of five centuries. Some of the objects are now lost; others are simply recycled building materials; still others are merely potsherds found in the repeatedly disturbed soil excavated from the city's streets. In sum, Cristofani does not present persuasive physical evidence of the existence of a Roman city. The one exception to the statement "few artifacts" is Lorenzo Ghiberti's fifteenth-century description of the early fourteenth-century discovery of a Lysippian Venus beneath the Malavolti palace. I will discuss Ghiberti's account in Chapter 3.
12. Cecchini, "Dove era la Siena romana?" 3. Weber, *Tabula Peutingeriana.*
13. Bortolotti, *Siena,* 5. Rizzitano, "Arabi in Italia," 106. Krueger, "Italian Cities and the Arabs, 51–2. Personal communication, Ludwin Paardekooper, Royal University of Groningen, The Netherlands. Hook and Bowsky have both stressed the significance of the belief. Hook, *Siena,* 6–7. Bowsky, "Siena," 277–282. For an examination of the ways in which the idea of the Middle Ages continues to inform aspects of Sienese life, see Dundes and Falassi, *Terra in Piazza.*
14. Schevill, *Siena,* 43, 300. From the *Diplomatico del monastero di S. Salvatore* in the Archivio di Stato di Siena, cited in Maroni, *Prime comunità cristiane,* 54. Cipolla, preface to *Banchieri e mercanti di Siena,* 9. Sterpos, *Comunicazioni stradali.* Sestan, "Siena avanti Montaperti," 46–46, note 40bis, citing Jung, "Itinerar des Erzbischofs. Maroni, *Prime comunità,* 54, also mentions Sigerico's itinerary, assigning it a date of 994. The German emperor passed through Siena on his way from Pavia to Rome, most likely in 1046. See Cipolla, *Banchieri,* 9. The oldest document in Siena's Episcopal Archives is Henry III's grant of the *Castellum vetus* to the bishop in 1055. The Sienese antiquarian Giugurta Tommasi wrote that at the siege of Antioch, Salimbeno Salimbeni, a member of the famous Sienese banking family, was the first over the wall. Bortolotti, *Siena,* 5 and 10, citing Tommasi's *Historie di Siena,* 1:126.
15. Construction on the second known cathedral began in 1189. Bortolotti, *Siena,* 10; and maps at 6, 27–28. Carli, *Il Duomo di Siena,* 2. Cristofani, *Siena,* 148. Balestracci and Piccinni, *Siena nel trecento,* 150–151 and map 9.

16. "Siena, centro popolato e dotato di mercati, di artigiani e di ricchezze." Idrisi, *Libro di Ruggero,* 101. Bortolotti, *Siena,* p. 10 and notes 30, 31 on p. 11; and fig. 14 on p. 27, "La cinta muraria alla prima metà del sec. XII," and note 119 on p. 55. Bowsky, *Medieval Italian Commune,* 19, gives 1328 as the year of peak population. Bowsky, "Black Death," 8, 9.
17. Van der Ploeg, *Art, Architecture, and Liturgy,* 162. Prager and Scaglia, *Mariano Taccola,* 8–9. Bowsky, "Siena," 281.
18. For more on the importance of horses in Siena, see Dundes and Falassi, *Terra in Piazza.*
19. Bowsky, *Finance,* 17. Bowsky, "Siena," 279. Caferro, *Mercenary Companies,* 129, citing Tortoli, "Per la storia della produzione laniera," 228, 231 for Siena. Giovanni Villani's figure of 100,00 for Florence and Hoshino's critique of Villani's figures in *L'arte della lana,* 194–206. Cipolla, *Banchieri.* For Spain see Basanoff, *Itinerario della carta,* 13. For Genoa and Fabriano see Irigoin, "Les origines de la fabrication," 62–67.
20. The purposes to which the wheels were put in Siena must have been obvious to those who recorded them because mills are never described with more than the generic terms for mill, *molendinum* or *molendina.* The laws governing mills appear in Sienese statutes right before or after the laws regulating the measure, milling, and transport of grain. See, for instance, the index of Ciampoli and Szabó's edition of the Sienese statutes regarding roads and fountains, called the Viarî, s.v. "molendina" and "molendinum;" published as *Viabilità e legislazione.*
21. Consiglio Generale Deliberazioni, vol. 48, f. 36, 5 August 1295, in Bargagli-Petrucci, *Fonti di Siena* 2:157–8.
22. White, *Medieval Technology and Social Change,* 86. Hills, *Power from Wind,* 16. See also Prager and Scaglia, *Mariano Taccola,* 126–126, 140–141. Prager and Scaglia, *Mariano Taccola,* 11, 37, 192, and I, 31v on p. 41. Taccola, *De ingeneis* I-II, 72v in Galluzzi, *Renaissance Engineers,* 27. *Dictionary of the Middle Ages* s.v. "Toledo." In an earlier work, Glick says, "Islamic, and later, Christian Spain became a new center for the diffusion of the *noria.*" He suggests that the giant noria at Toledo could have been a model for the first hydraulic wheel in Fez, Morocco. Glick, *Irrigation and Society,* 180–182.
23. Schevill, *Siena,* 22–24. Caferro, *Mercenary Companies,* xiv. Bortolotti, *Siena,* 13–14.
24. Paget says that Populonia extracted tens of thousands of kilograms of ore annually between the seventh and fourth century B.C.E. He illustrates his point with an image of a coin from the period bearing a sign of Populonia's trade: a hammer and tongs. Paget, *Central Italy,* 133 and figure 11. Cipolla, *Banchieri,* 11. *Enciclopedia italiana,* s.v. "Siena," says that Siena received Castello di Staggia and half the mine of Montieri. Cortese, *L'acqua, il grano, il ferro,* suggests that the Cistercians at San Galgano, twelve kilometers east of Montieri, had been involved in mining, which is not surprising since the

Cistercians had mines elsewhere in Europe. Braunfels reports the Cistercians were "the best agronomists, the best stockbreeders, and the best foresters of the later Middle Ages. They were experts in fisheries and the use of water, and the valued pioneers of mining and smelting," in *Monasteries,* 73. Bortolotti, *Siena,* 14. Schevill, *Siena,* 57. Bowsky gives the date 1185 in "Siena," 278, 279. Hook gives 1186 in *Siena,* 7. *Pirotechnia of Vannoccio Biringuccio,* xix.

NOTES TO CHAPTER THREE

1. "L'antico proverbio, che Siena il suo bello tien sotto a'piedi," in Tommasi, *Historie di Siena,* 3:170. Bargagli-Petrucci, *Fonti di Siena,*1:46, and Hook, *Siena,* 26, are two of the many historians who could not resist the allure of this quotation. Neither, however, cites a source. A sixteenth-century account of Charles V's visit mentions only a tour of the city; it is published in Vigo, *Carlo Quinto in Siena.* Sienese antiquary Agnolo Bardi, writing his "Seconda parte delle Istorie Senese," twenty years later in 1556, mentions only the Fonte Branda and a *passeggiata* (stroll).
2. Crouch, *Water Management,* 22.
3. Frontinus, *Aquaeductu* 92, 11.
4. Crouch, *Water Management,* 33, 59.
5. The ledgers for the years 1226–1229 are published as Direzione del R. Archivio di Stato di Siena, *Libri dell'entrata e dell'uscita.* Bortolotti, *Siena,* fig.16, p. 29. The Fonte Follonica would be enclosed during the next century, but the Fonte Ovile and the Fonte Peschaia remain outside the walls to this day. Bowsky uses a figure of fifty thousand people in 1328, in *Medieval Italian Commune,* 19. It is also possible that the slope of the aquiclude, the impermeable layer beneath the aquifer, tilts toward the south. A search of the literature revealed only national and regional surveys of Italian hydrology. The maps produced lack adequate detail for the area surrounding Siena. The most recent detailed study on the subject, Squyres, *Geology of Italy,* is typical of the broad-brush approach that often obscures local variations in land forms.
6. Neuburger describes wells 160–180 yards deep in Egypt, *Technical Arts and Sciences of the Ancients,* 418. Robins notes wells in Algeria "at depths varying from 120 to 1200 feet," *Story of Water Supply,* 26, 28. Of course, the documents could be biased toward recording details in institutional settings better than in domestic ones. See the list Bargagli-Petrucci gives in *Fonti di Siena,* 1:356. One of the only instances of a record of a domestic well or cistern I have seen was not even in the city of Siena, but in Montalcino, a town controlled by Siena. The document records the sale of a house to the hospital of S. Maria della Croce, along with "il diritto di prendere l'acqua da una cisterna di proprietà del venditore al prezzo di 30 fiorini," the right to take water from the seller's cistern. From the printed inventory of the Archivio Comunale di Montalcino, Fondo Ospedale di S. Maria della Croce,

Diplomatico, p. 133, no. 472. Courtesy of Ludwin Paardekooper, Royal University of Groningen.

7. "Con pozzi d'acque freschissime," in Boccaccio, *Decameron,* Branca, ed., Giornata 1, "Introduzione," line 90; "la freddissima acqua e in gran copia che quivi surgea," Giornata 3, "Introduzione," line 4.
8. Biccherna, vol. 35, Uscita, ff. 35–35v.
9. R. Archivio di Stato di Siena, *Costituto 1309–1310, distinzione* (section) III, rubric 115, 2:56–57. Surface water had likely become so contaminated with human and animal wastes that it is just as well that the paved surfaces diverted waste water to points below the city's wells.
10. Crouch, *Water Management,* 23–26. Allen provides excellent illustrations of how the trulli complexes in Apulia collected water in *Stone Shelters,* 107. Fabiano, "L'acqua," 373–386. Wycherley, *How the Greeks Built Cities,* 211.
11. Small, *Influent and Effluent,* 106. For the extraordinarily ornamental well heads of Venice, see Rizzi, *Vere da pozzo di Venezia.* There is a good diagram of a Venetian cistern in Braudel, *Structures,* 229. Baker gives a detailed description of the Venetian cisterns in *Quest,* 1:31–18.
12. *Costituto 1309–1310,* III, 137, 2:64. Like many others, this statute reiterated the provisions of an older one. The earlier version, from 1288, offers the same £15 incentive and "decem modios [*moggia*] calcine" for each cistern dug. Statuto 5, f. 192v. Edler indicates that a Florentine *moggio* was 24 bushels in the fifteenth century. The Sienese measure was likely similar. See her *Glossary,* 187. Bowsky confirms Edler's data, noting that the *moggio* equaled twenty-four *staia* and that a *staio* was roughly equivalent to a bushel. Bowsky, *Finance,* xx. "Item, xv libr. fratri Melano, operaio opere Sancte Marie, pro una citerna quam ipse fecit fieri in dicta opera, secundum formam Constituti Senarum," in Biccherna, uscita, vol. 37, f. 55, published in Bargagli-Petrucci, *Fonti di Siena* 2:125. *Costituto 1309–1310,* III, 138, 2:64. The year 1343 is the traditional date for the celebration of the first water to enter the Fonte Gaia. A lack of contemporary documents makes it difficult to pinpoint the year. For a discussion of the uncertainties see Hanson, *Fonte Gaia,* 7.
13. Nonetheless some Mediterranean communities were forced by their geography to drink water that had been stored for months on end. Robins gives some examples in *Water Supply,* 14. Today, Capri and Santorini supplement rainwater with imported supplies; until the 1940s households in Melos made rainwater last the whole year. Crouch, "Modern Insights," 95–96.
14. Bargagli-Petrucci's account of the General Council's discussion of the proposal can be found in *Fonti di Siena* 1:35–37; the document he cites to support it, the Deliberations of the Consiglio Generale, vol. 12, f.16v can be found in vol. 2, 126–127. See also Comune di Siena and Balestracci, *Bottini,* 17. Tito Giardi, an engineer and acquaintance of Bargagli-Petrucci, never published his study, except for the summary that Bargagli-Petrucci quotes in

Fonti di Siena 1:35–36. Giardi wrote that even with the advantages afforded by early twentieth-century engineering techniques, the Merse aqueduct would have been a difficult and costly project.

15. Bargagli-Petrucci, *Fonti di Siena* 1:219–224, 2:243. Domenico Battini, "Relazione intorno," 185–6. Balestracci, personal communication, November, 1993. Bowsky, "Black Death," 19–21.
16. Bowsky, "Black Death," 176–182, 185. Forbes argues that the Romans replaced the Greek siphon at Pergamon (originally built around 180 B.C.E.) with an aqueduct because aqueducts were more reliable. See Forbes, "Hydraulic Engineering and Sanitation," in Singer, 2:669.
17. Bortolotti, *Siena,* page 15, and note 13, citing Balestracci, "Approvvigionamento," 127–154. Information regarding the mills along the Merse River from p. 15, citing Zdekauer, *Vita privata,* 30.
18. Regulations mandating the construction of eighteen dry mills, six per district, appear in the Regolatori of 1453, f.216v-217. Their use in 1303 is recorded in *Cronache Senesi,* 83. "Item, x libr. quas habuerunt a magistro Ugolino Bernardini, solvente pro se et Ugolino Martini, consorte suo, pro pensione platee Comunis in qua est molendinum positum iuxta guazzatorium fontis Brandi, pro duobus annis conpletis in kal. januarii proxime preteritis, ad rationem c sol. pro quolibet anno." Biccherna, vol. 36, Entrata (revenue), f. 4., in Bargagli-Petrucci, *Fonti di Siena* 2:123.
19. Balestracci, "Terra dei mulini," 161.
20. The law ordered that along the road to Buonconvento "debiano ponere arbori ne le loro possessioni allato a la detta strada l'uno presso a l'altro der due braccia per conservamento de le greppe de la detta strada, et essi arbori debiano mantenere et potare et crescere competentenente." *Costituto 1309–1310,* III, 140, 2:65. For the effects of urbanization see Crouch, *Water Management,* 81; for an overview of environmental regulation beyond Siena, see Zupko and Laures, *Straws in the Wind.* Brooks, *Hydrology,* 75–81, 101–107. Goudie, *Human Impact,* 185–198, 251–253.
21. The exploitation of spring banks in this manner in the eastern Mediterranean and the Near East will be discussed further in Chapter 5.
22. Elsewhere in Italy, the word *bottino* has come to mean either a cesspool or a settling pool in an aqueduct. The one instance outside of Siena in which it is used to denote an underground aqueduct (though of which type is unclear) is in the legend of a 1784 map of the water supply of Todi, published in Bergamini, *Gli Etruschi,* fig. 93 p. 147, "Pianta ed alzata dei pozzi e bottini della fonte Cesia fatta per commissione del N. U. signor. . . ." For etymology see *Garzanti,* s.v. "bottino[2]." Bargagli-Petrucci says the first known use of the word *buctinus* is in 1226 in the first book of the Biccherna, which he provides in *Fonti di Siena* 1:34, 2:72. See Direzione del R. Archivio di Stato di Siena, *Libri dell'entrata e dell'uscita.*

23. The system of Roman aqueducts carved in the rock beneath Bologna is an exception, attributable in part to the nature of the rock. See *Acquedotto 2000.* "Ai bottini, più facilmente che alle fonti, possiamo riconoscere un principio romano." Bargagli-Petrucci, *Fonti di Siena* 1:33, 34, 40–41. He supports this assertion with a list of references to every Roman cuniculus and cistern he could find, from Aosta to Selinunte, citing numerous nineteenth-century digs and publications, mostly from the Accademia dei Lincei. However, he does not describe the similarities or differences at all. The major functional difference is that cuniculi were means of drainage, used as sewers or adits to remove water. Their presence in Rome seems to date from the Etruscan period.
24. "One reason why Roman ruins are so visually impressive is that the Romans were clumsy engineers: they had little sense of economy of means. . . ." White, "Flavor," 49.
25. Hodge uses the word "sinter" as a generic term for calcium deposits inside of aqueducts, as does Crouch. See Hodge, *Roman Aqueducts* and Crouch, *Water Management,* 69. In Sienese sources the word is either *gruma* or *tartaro;* the contemporary word is *calcaro.* Bargagli-Petrucci, *Fonti di Siena* 1:43, 2:362.
26. This injunction can be found in two contracts, from 1334 and 1339, reproduced in Bargagli-Petrucci, *Fonti di Siena* 2:214–216.
27. Quoted in Bargagli-Petrucci, *Fonti di Siena,* 1:218–219. Emphasis added.
28. Measurements confirmed by the author's own observations. The contract gives the dimensions as 1.5 *braccia* wide by three *braccia* high. I am using 0.6 meters to the *braccio.* Balestracci, for the sake of simplicity, uses an approximation of 0.5 meters to the *braccio.* American readers may find the equivalent of one *braccio* to two feet to be more intuitive. Hodge says Roman systems were also sized to accommodate construction and maintenance workers, not the flow of water, which used only a fraction of the tunnel's cross-section, as can be seen from the level of calcium deposits along the walls of aqueducts. Hodge, *Roman Aqueducts,* 94. Perhaps the plane at which the two surfaces meet is level, or slightly sloped to the south, in which case Bargagli-Petrucci's explanation may be correct. Riccardo Giacopelli, water supply engineer for the city of Siena, Ufficio Tecnico, Comune di Siena, confirmed Bargagli-Petrucci's explanation that the old bottini did indeed follow this principle, though a bit more loosely than suggested by Bargagli-Petrucci. Interview, July 1996. I am indebted to Anne Carey and her colleagues at the Department of Geology at the University of Nevada, Reno, and to Scott MacDonald at the University of Washington, Tacoma, for suggesting that clay deposits, such as the one under Siena, need not be level and that tilted or sloping impermeable layers can determine the ways in which water is shed from them.
29. Bargagli-Petrucci, *Fonti di Siena,* 1:36. Hodge estimates that the average minimum gradient in Roman construction was about 0.5 meters per kilome-

ter, or 0.5%. Hodge, *Roman Aqueducts,* 23. Frontinus suggests that the slopes were steeper in the older aqueducts perhaps "because the technique of leveling had not yet been precisely developed." *Aqueductu,* 18.

30. *Garzanti,* s.v. "spiraglio," confirms this usage. The etymology is from the verb "to breathe." Bargagli-Petrucci *Fonti di Siena,* 1:42, gives "smiragli" and "sboccatorii" as well, but they must be local usages as they do not appear in dictionaries of standard Italian; cf. Enrica Calchini, "Glossario dei termini tecnici nel Trattato I (Ms. Saluzziano 148) di Francesco di Giorgio," in *Prima di Leonardo,* 467. For a comparison between the two techniques see Goblot, *Qanats, 188–192.*
31. Robins, *Water Supply,* 97–98. Vitruvius describes the reservoirs, the primary function of which is distribution of water, Vitruvius, *On Architecture,* book 8, chapter 6, 2:183; his discussion of the benefits of letting sediment settle out is in the context of cisterns, not reservoirs, and can be found on p. 193. Frontinus, *Aqueductu,* 19. See also the caption to the reproduction of Wilbert's plan (figure 628) in Singer, 2:690. "Acciò che l'aqua venga chiara a la fonte del Campo e l'altri fonti. . . ." Registro of the Regolatori delle Ragioni, 26 May 1438, f. 87. Reprinted in Bargagli-Petrucci, *Fonti di Siena* 2:366–367.
32. Bargagli-Petrucci, *Fonti di Siena* 1:45, 235. Forbes, "Water Supply," 1:169.
33. Forbes, "Water Supply," 1:46. All lengths are approximate. They are derived from measuring the "Pianta degli acquedotti (Bottini) della città di Siena" published in the 1906 edition of Bargagli-Petrucci, *Fonti di Siena.*
34. Most of the channels leading to these wells have been blocked since the advent of pressurized running water earlier in this century.
35. Dimensions are taken from measured drawings published in Strack, *Brick Architecture,* plate 5, and figs. 6–8.
36. "Item lv sol. Magistro Iohanni, pro pretio leonis novi missi in fontem Brandum . . ." in Biccherna, December 1246, f. 23, published in Bargagli-Petrucci, *Fonti di Siena* 2:80.
37. Bargagli-Petrucci, *Fonti di Siena* 1:190–191.
38. The 1556 account is from "Memoria delli Bottini che sonno nella Città de Siena e che entrano in esse e fontane che fanno, e prima," published in Comini, *Rapporto,* 25–31. Courtesy of Ing. Riccardo Giacopelli, aqueduct engineer, and the Office of the Mayor, City of Siena. The second is from Tommasi, *Historie di Siena* 3:171–2.
39. Comini, *Rapporto,* 26.
40. Tommasi, *Historie di Siena* 3:171–172.
41. His familiarity with the semi-annual fee for a connection suggests he was a subscriber. Tommasi, *Historie di Siena* 3:172. Montaigne, *Journal,* 61.
42. *Costituto 1309–1310,* III, 97, 2:50.
43. Waley, *Siena,* 18. Braudel, *Structures,* 130. Prunai, "Notizie e documenti." Origo, "Domestic Enemy."

44. The Fonte Branda was the most abundant fountain in medieval and modern times. Because this comparative abundance was based on hydrological factors, we can infer that the nineteenth-century figures for maximum flow of water can give us a rough figure for an amount that other fountains did not exceed, in that century or earlier. Conversion of Domenico Battini's figure of 22,000 Sienese barrels per day by Cambi, *Del nuovo uso,* 7. Cambi refers to Battini, "Relazione intorno." "Precious Gallons," *Washington Post* 21 August 1999, Final Edition, p. A 13.
45. Hodge, *Roman Aqueducts,* 347–348. A detailed chronology can be found in Bargagli-Petrucci, *Fonti di Siena* 1:182–209.
46. Bowsky, *Nine,* 19.
47. For details about the Campo's evolution, see Guidoni, *Campo di Siena.*
48. Braunfels, *Urban Design,* 16. Braunfels is referring to the *Costituto 1309–1310,* III, 37, 2:29, Statuto 5 dist. III, Rubric 37. "Che in ciascuna casa., la quale si facesse di nuovo d'intorno al Campo del mercato, tutte le finestre si facciono a colonelli . . . 10 May 1297."
49. Braunfels, *Urban Design,* 61–67.
50. UNESCO added the center of Siena to its list of World Heritage Sites in 1995 confirming the judgment of many scholars as to the exceptional nature of the Campo.
51. Ghiberti, *Commentaries,* 165-166. According to Holt, Ghiberti wrote his account, "towards the end of his life," which would place it more than a century after the discovery of the Lysippian Venus during excavations beneath the Malavolti palace in 1325. Holt, *Documentary History,* 165. "Pro statua Fontis Campi. Item, quod statua marmorea ad presens in Fontem Campi posita, quam citius potest, tollatur ex inde, cum inhonestum videatur, et fiat ex inde, et de ea quod domini Duodecim videbitur et placedbit." *Concistoro, Deliberazioni,* f. 9, quoted in *Cronache Senesi,* 537–538.
52. Hanson discusses the various possible dates for della Quercia's birth, including 1371, 1374, and 1376 in *Fonte Gaia,* 10, 35, and document 1, p. 89.
53. For a succinct and sympathetic interpretation of della Quercia's program as an extension of the ideals of Ambrogio Lorenzetti's mural from the 1340s, "The Virtues of Good Government," in the Public Palace, see Hook, *Siena,* 99–100.
54. At the same time, the city moved the fountain about twenty meters to the west for reasons that remain unclear. Perhaps the restorers wanted to achieve a more symmetrical appearance by centering the fountain on the building behind it. See Guidoni, *Campo* and Hanson, *Fonte Gaia,* 1–2.
55. Bargagli-Petrucci, *Fonti di Siena* 1:227.
56. Bargagli-Petrucci, *Fonti di Siena* 1:226, 2:280. Schevill, *Siena,* 219–221.
57. Bowsky, *Finance,* 17–19, 43–45.

NOTES TO CHAPTER FOUR

1. These included the edition of 1337, the last major redaction of the statutes for 200 years. The next was to come in 1544. Bowsky, *Medieval Italian Commune,* 43. Magnusson makes this astute observation in *Water Technology in the Middle Ages,* 160–161.
2. Although sometimes translated as "constitution," the sense of "statuti," or "costituto," corresponds to a category of legal literature falling somewhere between city statutes and a state's constitution.
3. Grasso, *Cronaca Senese,* 537–538.
4. This, of course, is an oversimplified snapshot of a Siena's decline in political and economic primacy, a topic that is hotly debated and in no way settled.
5. "Anco, statuto et ordinato è, a belleza de la città . . . disfare . . . pozo o vero necessario," *Costituto 1309–1310,* III, rubric 265, 2:119. "Item, cum ultra Portam novam di Chamollia sit quedam porticciuola murata de terra iuxta alas murorum Comunis et occasione dicte porticciuole fiat ibi magnus lutus et etiam non potest videri pulcritudo Civitatis. . . ." Ciampoli and Szabó, *Viabilità e legislazione,* rubric 249, p. 183. The Viarî were assembled during the decade beginning in 1290. The Porta Camollia was used by southbound travelers arriving from Florence, Lucca, or France, and thus it warranted the special attention reflected in the statute. "Di fare rivedere et racconciare la strada Francesca . . . che va a Firenze . . . per onore et utilità del comune di Siena. . . ." *Costituto 1309–1310,* III, 304, 2:140.
6. "Et farò giurare le guardie de le fonti sopradette, coloro é quali contrafaranno, accusare: et a chi accusa, la metià [sic] de la pena sia conceduta." *Costituto 1309–1310,* III, 129, 2:61. Sienese building codes, which are among the earliest and most detailed in Europe, are discussed and published in Braunfels, *Mittelalterliche Stadtbaukunst,* 87–130; and idem, *Urban Design,* 61–67. Braunfels writes in *Urban Design,* "there is no comprehensive treatment of the urbanistics of Siena, although Siena is cited in all the authoritative works as an example" (p. 377, note 14). He adds that "a thorough study of all these ordinances and decrees, which goes further than the suggestions I make in *Mittelalterliche Stadtbaukunst,* 87 ff., has yet to be written" (p. 378, n. 15).
7. The best recent study of the frescoes is Starn, *Ambrogio Lorenzetti.* Starn's treatment of the paintings' content and context, illustrated by numerous details reproduced in color, makes the book especially useful to the student of urban history. Bruni, *Laudatio florentinae urbis,* 136–8. Alberti, *Art of Building,* 113.
8. A more detailed discussion of the shifting responsibilities for maintaining cities can be found in Ward-Perkins, *Classical Antiquity.* Goldthwaite, *Building of Renaissance Florence,* 69.
9. A brief description of each of the manuscript statutes can be found in Ministero dell'Interno, Pubblicazioni degli Archivi di Stato, *Archivio di Stato*

di Siena: Guida-Inventario. See especially 1:61–76. The process of publishing Sienese communal statutes (as opposed to statutes of specific guilds) began in the nineteenth century with Lodovico Zdekauer's edition of Statuto 2, published as *Constituto 1262* and Lisini's edition of Statuto 20, published as the *Costituto 1309–1310.* The publication of Sienese statutes continues in the 1990s with Ascheri's edition of *L'ultimo statuto,* and Ciampoli and Szabó's, *Viarî.* Zdekauer, "Sugli Statuti del Comune di Siena fino alla redazione del 1262," his introduction to *Constituto 1262,* lxxxxvii.

10. "De le mura [sic], fossi, porte, carbonaie, fonti, ponti, vie et strade del comune di Siena." *Costituto 1309–1310,* III, 1, 2:15–16. The first law of the third section of Statuto 2, from about forty years earlier, begins in a similar way, but the title of the law is more narrow, mentioning only the walls and openings through it: "De muro comunis et portis et porticciolis." Zdekauer, *Constituto 1262,* III, 1, p. 275. The wording of each law is about the same, with the details of enforcement more clearly laid out in the 1309 version. Bowsky defines the *giudice sindaco* as the guardian of the constitution. See his *Nine,* 42, and idem, "Constitution and Administration." I have deferred to Bowsky's nomenclature for all translations of city offices and jurisdictions. The secrecy the regulations enjoined upon inspectors may have been intended to free them from threats of intimidation and graft. Florence had a similar system of enforcement. The city employed guild agents called *ricercatore* (searchers) to enforce building codes. See Goldthwaite, *Building,* 210. Et lo giudice sindaco del comune di Siena sia tenuto et debia andare et rivedere le mura . . . segretamente, secondo che mellio fare si potrà . . . Et debia denuntiare a missere lo capitano del comune et del popolo di Siena. Et missere lo capitano predetto debia credere a la relatione sua . . . Et se 'l predetto giudice sindaco le predette cose non farà, perda del suo salario XXV. libre di denari, le quali el camarlèngo et iiij. proveditori sieno tenuti del salario del detto sindaco ritenere per lo comune di Siena." *Costituto 1309–1310,* III, 1, 2:15–16.
11. Sienese legal scholar Donatella Ciampoli has produced a pioneering attempt to collate one statute with the General Council deliberations that determined its final form in her article, "Una raccolta di provvisioni senesi agli albori del XV secolo: il 'libro della catena.'"
12. For more on the office of the Podestà and other officials in Siena, see Bowsky, *Nine,* 23–34. For a comparison of the government of Siena and that of other Tuscan cities, see Waley, *Italian City-Republics.*
13. "Che la podestà sia tenuto avere uno maestro di pietra et lo manovale, per acconciare le vie . . . uno buono maestro di pietra et uno sufficiente manovale . . . excetti li dì solenni et festarecci." *Costituto 1309–1310,* III, 8, 2:18. "Et . . . operarius religiosus Operis Sancte Marie eas debeat revidere. . . ." *Viarî,* 95, p. 115. The Cathedral Works were also known as the Opera Sancte Marie. "De tribus eligendis super revidendis fossis et carbonariis. . . ." in

Zdekauer, *Constituto 1262,* 281; Statuto 11, dist III, f. 158. The statute reappears in Statuto 20 as: "Anco . . . tre huomini savi et discreti uno di ciascuno Terzo, é quali ciascuno mese, debiano andare dintorno a li fossi . . . et revedere. . . ." *Costituto 1309–1310,* III, 19, 2:22. "Anco . . . due buoni huomini di ciascuno Terzo de la città, é quali con maestri savi. . . ." *Costituto 1309–1310,* III, 25, 2:25.

14. "De la electione di coloro é quali debono rivedere la fonte a Follonica . . . IIII. buoni huomini del Terzo di Sancto Martino, é quali per saramento fatto, sieno tenuti andare a la fonte a Follonica et vedere et sottilmente investigare et invenire, ine o vero in quale luogo mellio et più utilmente et più convenevolmente el lavatoio de la fonte predetta. . . ." *Costituto 1309–1310,* III, 72, 2:41. Cf. Statuto 4 (1282–89), f. 16, in Bargagli-Petrucci, *Fonti di Siena* 2:9–10. Bowsky, *Nine,* 14. "Et . . . elegere X. buoni huomini et discreti; cioè, due del popolo di Sancto Pietro ad Ovile et due del popolo de l'abbadia di Sancto Donato et due del popolo di Sancto Andrea et con essi debiano attendere et rivedere et ordinare in quale luogo più utilmente et più agevolmente per lo comune di Siena si debia e possa fare una fonte dentro a la porta del Piano d'Uvile. . . ." *Costituto 1309–1310,* III, 115, 2:56–57. Balestracci and Piccinni provide a map showing the borders of fourteenth-century *popoli.* See Balestracci and Piccinni, *Siena nel Trecento,* Carta 1, "I popoli di Siena attraverso I dati della "Tavola delle Possessioni" (1318–1320)." Why the numbers do not add up to ten is not clear. Was there an at-large member? "Et . . . farò guardare, mondare, et racconciare a la rinchiesta di IIII. o vero di più huomini de la contrada, 've la fonte sarà, a l'expese del comune di Siena. . . ." *Costituto 1309–1310,* III, 126, 2:60–61.
15. Moorman, *History of the Franciscan Order,* 40–45, 216–217. "De viis magistris circa Civitatem Senarum rubrica . . . eligantur . . . duo religiosi viri boni et discreti . . . qui . . . teneantur [et debeant] totam stratam Francigenam. . . ." *Viarî,* 101, p. 120.
16. These distinctions are illustrated in Alberti, *Art of Building,* 331, which will be discussed below. Balestracci and Piccinni, *Siena nel Trecento,* 145.
17. "Item cum in stratis et viis publicis aquarum copia sit necessaria et utilis valde et in via, qua itur a[d] Montalbuccium . . . uno fonte fiendo. . . ." *Viarî,* 36, p. 89. See also rubrics 237 and 177.
18. Cortese catalogs some of these mills in *L'acqua, Il Grano, Il Ferro,* 225–316.
19. Bowsky, *Finance,* 17.
20. Consiglio Generale Deliberazioni, vol. 196, f. 129, 16 December 1389, in Bargagli-Petrucci, *Le fonti di Siena* 2:279–280. Balestracci and Piccinni cite this document and an earlier discussion in the same volume of deliberations to support their assertion that the overflow from the fountains supplied "an imprecise number of mills outside the walls." See *Siena nel Trecento,* 163. Their map of workshops and points of sale, "Carta n. 10—Punti di Produzione e di vendita nel corso del xiv secolo," indicates eight mills out-

side the walls from the Fonte Branda. Braudel, "Daily Bread," 104–145 in *Structures of Everyday Life.* See also Mumford, "Daily Grind," in *Technics and Human Development,* 137–139.

21. The first publication of such a law was in 1262. "De fossatis occupatis . . . per que aqua currere consuevit, et aqua impediatur ire per cursum suum et meatum anticum. . . ." Zdekauer, *Constituto 1262,* III, 248, p. 350. It is reiterated with the title "De non occupandis fossatis" in Statuto 5 (1288–1293), III, f. 193; and again in 1309 as "Di non occupare li fossati. Et qualunque occuparà li fossati per li quali l'acque correre solevano et l'aqua sia impedita andare per lo corso, et questo noccia et nuocere possa a le terre o vero molina di sopra. . . ." *Costituto 1309–1310,* III, 141, 2:65. "Che uno muro per traverso si faccia nel contorno de la meza parte del bagno a Vignone si che 'l bagno, el quale si murarà, non si voiti per sè quando macinaranno le molina. . . ." *Costituto 1309–1310,* III, 175, 2:81. The law first appeared as "De muro fiendo in certa parte balnei de Vignone." Zdekauer, *Constituto 1262,* III, 277, p. 360. "Che neuno possa o vero debbia da la fonte di val di Montone tenere o vero fare alcuna fossa o vero steccata per aqua ritenere o vero colliere, a CC. braccia. . . ." *Costituto 1309–1310,* III, 262, 2:116. The statute says four *braccia* (about two meters). "Et . . . che a le sue expese facciano o vero facciano fare ponte di pietre o vero mattoni o vero legname di quercia, o vero con travi buone et ottime, sì che li uomini et le bestie possano bene passare." *Costituto 1309–1310,* III, 94, 2:49. This law seems to be a reiteration of a statute from twenty years earlier, "De recidentibus vias per goris molendinorum," Statuto 5 (1288–1293), III, f. 183; which in turn seems to be an amplification and expansion of a law ordering the construction of a single bridge: "Item cum pons qui est super goram molendini. . . ." Statuto 5 (1288–1293), III, f. 182.

22. "De pena rumpentis goram vel stechatam molendini . . . xxxv. pro qualibet vice. Zdekauer, *Constituto 1262,* III, 252, p. 351. At the end of the century it was reiterated with the title, "De pena frangentis goram vel stecchati molendini," in Statuto 5 (1288–1293), III, f. 194. In 1309 the law was still on the books: "Et . . . alcuno danno farà, sia punito, per ciascuna volta, in XXV. libre di denari." *Costituto 1309–1310,* III, 146, 2:67. After the shortage of grain following the warm winter and drought in 1303 the city renewed its efforts to ensure a steady supply of grain by buying the port of Talamone. See *Cronaca senese dall'anno 1202 al 1362,* 83–84. Among the mill owners there seem to have been some of the very members of the rural nobility over whom the city was attempting to gain greater control. The ongoing struggle to rein in rural magnates could have provided another motivation to regulate their milling operations so closely.

23. Cortese, *L'acqua, Il Grano, Il Ferro,* provides maps on 33, 137, 196, 225. "Quod nullum hedificium fiat in fluminibus iurisdictionis et senarum, propter quod impediatur molendinum vicini. Et in aqua de Riluogo et

Bulgione et de Bocçone et Sorra et Fusola et Luco et Umbrone et Tressa et Rosario et Malena, et in aliis aquis nostre iurisdictionis et districtus, et Arbia a Pancole usque ad Umbronem, nullum hedificium nec aliud fieri permittam [sic]; et si factum est, illud destrui faciam, propter quod suus vicinus perdat vel dissipetur suum molendinum factum vel prius inceptum, ita quod superiora molendina non impediantur molere per inferiora et e contra." *Constituto 1262,* III, 249, p. 350. They are repeated in 1288–93: "De molendinis non construendis. Et in aqua de Riluogo et Bulgione et Sorra et Luco et Umbrone et Tressa et Rosario Malena et alius aquis iurisdictionis et districtus Sen. et Arbia et a Pancole usque ad Umbronem nullum hedifitium fieri permictam et si factum et illud destrum faciam prop quod vincinus suus persdat vel dissipetur suum molendinum factum vel prius inceptum ita quod superiora molendina non impediantur molere per inferiora et contrario." Statuto 5, dist. III, ff. 193–194v; where several related statutes appear, including "De novitatibus factis in molendino dissipandis," ff. 193–193v. "Et ne l'aqua di Riluogo, et Bolgione [sic] et Sorra et Fusola et Luco et Ombrone et Tressa et Rosaio et Malena et ne l'altre aque de la giurisditione et distretto di Siena et Arbia, et da Pancole infino ad Ombrone, neuno hedificio fare lassarò, et, se fatto è, esso disfare farò . . . sì che le molina di sopra non sieno impedite macinare per quelle di sotto, nè quelle di sotto per quelle di sopra." *Costituto 1309–1310,* III, 142, 2:65–66.

24. "Anco, statuimo et ordiniamo che missere la podestà di Siena sia tenuto et debia, sotto pena di C. libre di denari del suo salario, fare terminare, infra IIII. mesi de l'entrata del suo regimento, per due buoni huomini per ciascuno Terzo, é quali s'elegano per li iiij. proveditori, tutte et ciascune steccate de le molina del contado et giurisditione di Siena. La quale terminagione fatta, si reduca in scrittura per scrittura autentica, per lo comune di Siena. Et neuno debia, nè possa, fatta la terminagione predetta, essa alzare o vero mutare in alcuna cosa, a pena di C. libre di denari per ciascuno et ciascuna volta." *Costituto 1309–1310,* III, 143, 2:66. "Et se alcuna novità fatta fusse in alcuno molino, fatto dipo l'altro, et quella novità impedisse el prima fatto macinare, quella novità disfare farò. . . ." The £100 fine for failing to undo such changes was severe, at least when compared to the 10s. exacted from those who threw offal into the fountain of the Val di Montone. *Costituto 1309–1310,* III, 144, 2:66–67; "Che non gitti sozura sopra la fonte di val di Montone." Ibid., 60.

25. "Et questo capitolo abia luogo per li laici et intra li laici, et non per li cherici o vero religiosi; salvo che se li cherici o vero religiose persone vorranno et consentiranno che le loro molina et steccate si debiano terminare. . . ." *Costituto 1309–1310,* III, 144, 2:66–67. For example, "Anco statuimo et ordiniamo che se lo spedale Sancte Marie of vero el rectore d'esso, per esso spedale, vorrà hedificare o vero trattare alcuno molino d'esso spedale di val d'Orcia, nel luogo nel quale è posto, che sia licito questo fare senza contra-

dictione d'alcuno." *Costituto 1309–1310,* III, 152, 2:70–71. Cortese discusses ecclesiastical and monastic expertise and investment in mills, *L'acqua, Il Grano, Il Ferro,* 111–117.

26. "Di costregnere li vicini vendere la terra a chi volesse fare molina." *Costituto 1309–1310,* III, 151, 2:70. Clerical and monastic landowners were exempt in so far as the bishop's permission was needed to carry out these provisions against their holdings. A similar law, augmented with the justification that it was imposed "for the great and evident utility of the people and the city" ("pro magna et evidenti utilitate populi et comunis Senarum"), was already on the books in 1262. The earlier version of the law lacked the exemption for religious owners. *Constituto 1262,* III, 257, p. 353.
27. A history of the mechanization of Siena's industries and their transfer to mill sites outside the city has only begun to be written. See for instance Cortese *L'acqua, il grano, il ferro.*
28. "Steps in the Manufacture of Woolen Cloth in Italy" in Edler, *Glossary,* 324–329. The rate and extent to which these processes had moved out of households to centralized facilities in fourteenth-century Siena is not clear from the available documents and secondary sources. That a significant portion did take place at centralized facilities can be inferred from the involvement of the city and guilds in the construction and regulation of basins for fulling and dying. Carus-Wilson found that the guilds generally built their own infrastructure. See "Woolen Industry," in *Trade and Industry,* 656. Magnusson, *Water Technology in the Middle Ages,* 154–155. Banchi and Polidori edited and published many of these statutes in the series *Statuti senesi.*
29. "Item statuimo et ordinamo, che neuno sottoposto dell'Arte de la Lana di Siena possa nè debbia lavare ne la piscina di sopra alcuna lana nè stame nè panni, se no' lana o stame o panni vermelli e verdi e gialli e sanguegni e uricellate e ranci, le lane e stame alluminate, bambagini alluminati nei dicti colori. Possansi anco lavare ne le decta piscina panni sodi e crudi, e tutti panni di lana d' ogne colore; esceptati panni tenti in guado e in nero, e panni lani e bambagini tenti in indico, e tucti panni e bambagini infolliati, ei quali in essa piscina lavare non si possano. Ne la seconda piscina si possa lavare lana e stame di guado, e panni che si tegnessero in guado, e tutti colori di guado; e non neuno altro colore. Ne la terza piscina si possa lavare lana sucida di garbo e nera, e panni neri e di ciascuno colore di nero e indico, e panni e bambagini infolliati, e telecte nere. Ne la piscina nuova di sotto si possa lavare lana sucida grossa, e méttarvisi pelli e coiame in mollo; e non in neuna altra; e tutto el lavorío el quale si pertiene all' arte del coiame." In *Statuto dell'Arte di Lana,* 270, with definitions from the glossary published in the same volume, supplemented by definitions from De Mauro, *Grande Dizionario* and Edler, *Glossary.*
30. Carus-Wilson, "Woolen Industry," 616.

31. "Con ciò sia cosa che sia molto utile per andare a le nostre gualchiere." *Statuto dell'Arte di Lana,* 187. "Item . . . mandare due buoni uomini di questa Arte sottoposti secretamente, a le gualchiere dell'Arte, due volti in sei mesi del loro consolato; cioè una volta nei primi tre mesi, e l'altra nei tre mesi seguenti: et debbiano provedere se le cose dell'Arte sono mantenute in stato, secondo che ne la promissione de le carte si contiene." *Statuto dell'Arte di Lana,* 188, 249–259.
32. "Et la detta aqua soperchia, nel detto bottino di fonte Branda, acciò che li consoli de l'arte de la Lana possano avere magiore abondanza d'aqua." *Costituto 1309–1310,* III, 113, 2:56. "Et non lassarò lavare lana o vero mettere cuoia nel lavatoio et truoghi et guazatoi di fonte Branda. . . ." *Costituto 1309–1310,* III, 107, 2:53. The word can mean either a pool through which one walks livestock, especially horses, or waters livestock, or a pool for washing wool. "Che sia licito a li tegnitori fare bottino ne le vie pubbliche per derivare la loro aqua . . . d'ampieza d'uno mattone, per lo più, coperto di mattoni. . . ." *Costituto 1309–1310,* III, 135. 2:63.
33. This location is Bargagli-Petrucci's conjecture. The general patterns of continuity of land use that prevail elsewhere in Siena support his judgment. See his chapter on the fountain of Vetrice and the map of fountain locations. Recent scholarship has not yet addressed the location of the Fonte Vetrice. New tools, such as ground-penetrating radar, may soon allow archaeologists to locate and document the abandoned network of aqueducts that once supplied the fountain without disturbing the streets or buildings that have posed obstacles to excavations. After the Fonte Vetrice was abandoned for unrecorded reasons around 1306, both functions were consolidated around the Fonte Branda. Bargagli-Petrucci, *Fonti di Siena,* 1:337–344. It is possible that the fountain dried up as a result of either the expansion of the Fonte Branda system, which could have diverted water from the Fonte Vetrice, or that the growth of the city "upstream" and consequent building and paving diverted surface water from the latter fountain's recharge area. Without a precise location for the fountain it is difficult to do more than speculate as to the exact cause.
34. Bowsky, *Nine,* 205–212.
35. "Et non lassarò ne la via de la fonte a la Vetrice, da la porta del comune infino a la detta fonte, o vero altre fonti, in alcuna parte a C. braccia, gittare interami o vero cose putride, o vero scorticare alcuna bestia o vero carname, o vero terra gittare." *Costituto 1309–1310,* III, 67, 2:39. "Et neuno debia gittare o vero fare gittare alcuna sozura o vero puzura, di dì o di notte, da la fonte di val di Montone in su per alcuno modo o vero carname o vero sangue." *Costituto 1309–1310,* III, 124, 2:60. The phrase "from above" refers to the vulnerable position of most fountains: built into a hillside. "Et la fonte di val di Montone si debia guardare a le spese del comune di Siena. . . ." *Costituto 1309–1310,* III, 123 2:60.

36. Ancho, iii libr. a Biagio Brandi, guardia di fonte Branda.
Ancho, x sol. a Chambio, guardia di fonte a Follonicha.
Ancho, x sol a Gíovanino, guardia di fonte a Uvile.
Ancho, x sol. a Gualfreduccio, guardia di fonte a Pescharia.
Ancho, x sol. a Bernarduccio Palmieri, guardia di de la fonte di Val di Montone
Ancho, x sol. a Lolo Mini, guardia di fonte di Vallerozi.
Ancho, x sol. a Mino Muschio, guardia di fonte di fosso da' frati Romiti.
Biccherna, Uscita, vol 125, f. 277, published in Bargagli-Petrucci, *Fonti di Siena* 2:182.
37. Gabbitas, it should be noted, seems to use only sources from medieval England. Gabbitas, "Leather and Leatherworking," 530–532. Waterer provides some useful illustrations by Jost Amman and from Diderot's *Encyclopédie* in "Leather," 151–153. Forbes, "Leather in Antiquity," 1–9. Welsh, *Tanning in the United States.* Wulff describes traditional tanning methods in *Traditional Crafts of Persia,* 230–232. "Anco . . . che neuno gitti o vero gittare facia alcuna calcina con pelo et senza pelo, o vero alcuno calcinaccio o vero pelo nel guazatoio di fonte Branda. . . ." *Costituto 1309–1310,* III, 108, 2:54. "Et non . . . mettere cuoia nel lavatoio et truoghi et guazatoi di fonte Branda. . . ." *Costituto 1309–1310,* III, 107, 2:53. "Di non pelare o vero scorticare cuoia ne le cantine nè apo esse, le quali sono nel piano di fonte Branda . . . neuno debia scuoiare o vero in macero o vero concia mettere o vero fare mettere alcune cuoia o vero coiame, nè pelare o vero fare pelare." *Costituto 1309–1310,* III, 109, 2:54.
38. Goldthwaite, *Building,* 1, 2, 8, 19, 174. General Council of the Bell, "Ordinamenta supra igne extinguendo," 4 February 1282, published in Bargagli-Petrucci, *Fonti di Siena,* 2:6. A discussion of firefighting can be found in the chapter Balestracci and Piccinni devote it in *Siena nel Trecento.* Tura del Grasso, *Cronaca Senese,* 265, 297.
39. Alberti, *Art of Building,* 331.
40. Alberti, *Art of Building,* 13–16. Somewhat surprising perhaps is the lack of any sanction, analogous to the ones addressing livestock, preventing people with visible symptoms of disease from drinking at public fountains.
41. "Di non mettere o vero lavare lana o vero cuoia ne li lavatoi di fonte Branda." *Costituto 1309–1310,* III, 107, 2:53. Statuto 1282–89, f. 41, in Bargagli-Petrucci, *Fonti di Siena* 2:7.
42. One of the few studies touching upon early-modern laundry is by Peter Stallybrass. He emphasizes the extraordinary longevity of the materials used to construct garments. Using probate records, pawn shop account books, and theater inventories, Stallybrass learned that sturdy fabrics could survive not only recycling through several garments but also could outlast the lives of several owners. "Worn Worlds." The *rigattieri,* or used clothing dealers, were among the *arti medie* (middle guilds) who shared some political power in thirteenth-century Florence. See Schevill, *History of Florence,* 154, and Edler,

Glossary, 242–243. Although the trade in textiles was international, extreme caution must be used in inferring anything from later, English documents. The paucity of documents is reflected in studies by historians of the early-modern era such as Braudel, who supports his discussions of sanitary conditions and bathing in *Structures* 556, 330, by citing Mumford, *City in History,* 296ff., 468, who in turn cites Thorndike's classic, "Sanitation, Baths, and Street Cleaning." Forbes, "Washing, Bleaching, and Felting," 81–89, discusses the specialization of fulling and laundering in Ancient Egypt and Rome but offers no data on medieval practice. The only study on premodern laundry in Italy that I have been able to locate is an exhibition catalog published by the local Tourism Commission of Grassina, a village that was once famous as a place where residents of Florence sent their laundry. See Guerrini, *Maria lavava.* Guerrini's illustrations of laundry drying in the sun recall those of the Sienese fountains published in Bargagli-Petrucci, *Fonti di Siena.*

43. Roaf, "Settlement Form and Qanat Routes," 60. For some Sienese examples, see Balestracci, "Regulation of Public Health."
44. The reasons for parallels may go even deeper, an issue addressed in chapter 5. Crouch, *Water Management,* 22.
45. These documents could have been a target of the arsonists who destroyed city records in 1355. I have not seen any reference to court records in any history of Siena nor in any finding aid to the Archivio di Stato. The archivists at the Archivio di Stato, in response to several requests in 1993–1994, were unable to locate any records of criminal proceedings for the period in question.
46. Mumford, *City in History,* 287, 288–293.
47. Alberti, *Art of Building,* 113.
48. Roaf, "Settlement Form and Qanat Routes," 58–60. Crouch makes a similar point in her essay, "Modern Insights," 99–101.

NOTES TO CHAPTER FIVE

1. Ward-Perkins gives a definitive description of *cuniculi* and their impact on the landscape in his *Landscape and History,* 13. See also Judson and Kahane, "Underground Drainageways," 74–99 and Figure 30. For Florence, see Chiostri, *L'acquedotto romano.* For Bologna see *Acquedotto 2000.* For adits, see Klaubert, "Qanats," and Goblot, *Qanats,* 185. Neither Klaubert nor Goblot has been able to date any of the Bavarian or Bohemian qanats before 1750, and many of them have remained undated since their first discovery in 1964. For sappers, see Pepper and Adams, *Firearms and Fortifications,* 104–5. For the relationship between adits and cuniculi see *Dictionnaire des antiquités grecques et romains,* s.v. "Cuniculus." For adits as a system of water supply in Jerusalem and nearby sites, including a comparison of adits and qanats in that setting, see Ron, "Qanats and Spring Flow Tunnels." For an eleventh-century treatise on qanats see Mohammad ibn al-Husayn al Karagi, *Civilisation des eaux cachées.* See also Wulff, "Qanats of Iran," Goblot,

Qanats, and English, "Origin and Spread of Qanats." For the relationship between adits and qanats, see Forbes, "Hydraulic Engineering, 666. For Spain and parallels between the diffusion of qanats and norias, see Glick, *Irrigation and Society,* 180–187.

2. Glick, *Islamic and Christian Spain,* 226–227. Needham, *Introductory Orientations.* White, "Tibet, India, Malaya." Glick, *Islamic and Christian Spain,* 217–220.
3. Kroeber, *Anthropology,* 195. Kroeber describes the concept in his essay, "Stimulus Diffusion." For an overview of the growing literature of technological diffusion, see Rogers, *Diffusion of Innovations.* Beals noted the difficulty of evaluating Kroeber's theories "so soon after his death," but argued for widespread impact, especially on archaeologists. Beals, "Kroeber, Alfred L."
4. White, "Study of Medieval Technology," 521. Kroeber, *Anthropology.*
5. Needham, *Introductory Orientations,* 244.
6. For instance, Watson begins his study on the diffusion of Islamic agriculture without a discussion of the diffusion model or citing its originator. Watson, *Agricultural Innovation.*
7. Botanists also use the words "introduction," "transmission," and "assimilation" to describe the diffusion process. See for instance, Bermejo, "Economic Botany and Ethnobotany in Al-Andalus." White, "Indic Elements."
8. Cassirer, *Renaissance Philosophy of Man,* 142. White, "Eurasian Context of Medieval Europe," 1–2.
9. Lindberg, *Beginnings of Western Science,* 170. Hassan and Hill, *Islamic Technology,* 24, 26, 32.
10. Amari, *Biblioteca arabo-sicula. Encyclopaedia of Islam.*
11. Singer, *Mediterranean Civilizations,* 755.
12. White, review of Singer, *Mediterranean Civilizations* 131. De Solla Price, "On the Origin of Clockwork;" idem, *Gears from the Greeks.* The best summary of White's later work on this subject is his "Eurasian Context."
13. Hassan and Hill, *Islamic Technology,* 32–33. Pryor, *Geography, Technology, and War,* 135. Udovitch, *Partnership and Profit,* 3. Constable, *Medieval Iberia;* idem, *Trade and Traders in Muslim Spain.* Singer, "Epilogue," 773. Goitein, *Mediterranean Society.* Glick, *Islamic and Christian Spain,* 235–247; idem, "Moriscos and Marranos."
14. Strayer, *Dictionary of the Middle Ages.* White, "Technology, Western."
15. Taccola, *De machinis.* Anonymous of the Hussite Wars, *Technological Illustrations.* Giorgio Martini, *Trattati di architettura.* White had previously noted the lack of an edition of Giovanni de' Dondi's astronomical clock manuscript, but that was remedied with the publication of Giovanni de' Dondi, *Planetarium* and Bedini and Maddison, *Mechanical Universe.* See White, *Medieval Technology,* 126, and idem, "Flavor," 37–38.
16. Hassan and Hill, *Islamic Technology,* 12, 15, 21–22, 109–110, 143. al-Jazari, *Book of Knowledge.* The only modern edition of any of Fontana's ten works is

Macchine Cifrate. Compare the illustrations reproduced by Prager with those in Clagett, "Life and Works of Giovanni Fontana;" and Fontana, *Macchine Cifrate* with those in al-Jazari, *Book of Knowledge.* Fontana, *Macchine Cifrate,* 161. Prager, "Fontana," 346. Thorndike, "Giovanni da Fontana," remains essential to any study of Fontana.

17. Haskins, *Renaissance.* Heyd, *Histoire du commerce,* 1:180. Haskins, *Renaissance,* 63–64, 278–291.
18. Pegolotti, *Practica,* 360–383. Evans estimates the text to date from 1310–1340, xv. Evans provides a useful glossary on pp. 411–435. A convenient, brief English translation of the 288 names of the spices on this list, along with some useful editorial comments on Evans's translations, can be found in Lopez and Raymond, *Medieval Trade,* 108–114. Edler, *Glossary.*
19. Singer, "Epilogue," 770–771. White, "Study of Medieval Technology," 525.
20. Goitein, *Mediterranean Society,* 1:45–46. Udovitch, *Partnership and Profit,* 170–176. Watt, *Influence of Islam.* Edmund Wiener, Deputy Chief Editor, *OED2,* Personal communication, 22 July 1998.
21. Hassan and Hill, *Islamic Technology.* Menocal, "And How 'Western' Was the Rest of Medieval Europe?" 183. For Menocal's argument, see her *Arabic Role.*
22. White, "Cultural Climates,"183–184, and some of the work mentioned above. Kealey, *Harvesting the Air.* Hills, *Power from Wind,* 16, cites Taccola's notebook as the first instance in Europe. See Langdon, "Review," and Coutant, "L'étude des moulins à vent médiévaux."
23. For a discussion of cannon, silk, magnetic compass, sugar, paper, and printing see White, "Eurasian Context." For troubadour poetry see Menocal, *Arabic Role,* 71–93. For elements of mechanical clockwork see Hill, *Islamic Science,* 229–231. Lopez, "European Merchants in the Medieval Indies," 166. Yule, *Cathay and the Way Thither.* Dawson, *Mongol Mission.*
24. Heyd, *Histoire du commerce,* 1:180. Goitein, *Mediterranean Society,* Goitein, *Mediterranean Society,* 1:50–51, 59, 62, 66, 70, 4:190,192. Bentley, *Old World Encounters,* 27, 111–116.
25. "Benuccio [di Giovanni Salimbeni] e pagato d'argento, e di rame e di scorze e di denari. . . ." in Tura del Grasso, *Cronaca Senese,* 521. Surprisingly little has been published on Sienese merchants. See Zdekauer, *Mercante senese nel Dugento.* Astuti, *Libro dell' entrata e dell' uscita.* Cardini, *Banchieri e mercanti di Siena.* The phrase "inviting sea" (from *pontos euxeinos,* literally inviting to foreigners or hospitable) is from Goitein, *Mediterranean Society,* 4:43.
26. Pryor, *Geography,* 3, 36, 51–52. Heyd, *Histoire du commerce,* 1:181–182 (italics are Heyd's).
27. Lopez, "European Merchants," 172. Pegolotti, *Practica,* 22.
28. Goitein, *Mediterranean Society,* 4:69. Boccaccio, *Decameron,* Musa, trans., VIII:10, 544–545. Arabisms in parentheses from *Decameron,* ed. Branca, 1009. Branca notes that this particular passage provides important evidence for the early use of the modern system of "docks and warrants."

29. The reader will note that each of the nouns in this sentence has an Arabic etymon.
30. Pegolotti, *Practica,* 360–383. Lopez and Raymond explain the distinction in *Medieval Trade,* 108. Tangheroni, "Fibonacci, Pisa, e il Mediterraneo," 22.
31. Menocal, *Arabic Role,* 58–60.
32. For example, see Carus-Wilson, "Woolen Industry. " Lopez, "Silk Industry in the Byzantine Empire. " Mazzaoui, "Cotton Industry. " Wulff, *Traditional Crafts of Persia.* Hoshino, *L'arte della lana a Firenze;* idem, "Rise of the Florentine Woollen Industry. " Tortoli, "Per la storia della produzione laniera a Siena."
33. White, "Cultural Climates," 183–184.
34. Hassan and Hill, *Islamic Technology,* 53–54. Although Hassan and Hill assert that the first use of the floating mill was in the eleventh century, Singer notes that such mills did appear briefly in sixth-century Rome, where they provided an improvised solution to the need for water-powered grain mills after aqueducts that had powered the flour mills were severed. Cited in Prager and Scaglia, *Mariano Taccola,* 68. White, "Review of *Mediterranean Civilizations,*" 132.
35. White, *Medieval Technology and Social Change.* Watson, "Arab Agricultural Revolution," idem, *Agricultural Innovation.* Long, *Technology and Society,* 33.
36. White, "Review of *Mediterranean Civilizations,*" 131. Pegolotti, *Practica,* 367–368, 400. Hassan and Hill, *Islamic Technology,* 233–236.
37. Hill, *Islamic Science,* 185. Watson, *Agricultural Innovation.* Watson's monumental study needs to be emulated for France, Italy, and Greece.
38. Hersey, "Water-Works and Water-Play in Renaissance Naples." Hassan and Hill, *Islamic Technology,* 60. Hill, *Islamic Science,* 229. Hassan and Hill, *Islamic Technology,* 6, 55, 60.
39. Hassan and Hill, *Islamic Technology,* 33, 152–153, 190–191.
40. Irigoin, "Origines," 66. Basanoff, *Itinerario della carta,* 25. The Museo della Carta in Fabriano suggests a direct transmission from Damascus to Fabriano. http://www.museodellacarta.com/ accessed 3 May 2004.
41. *Dictionary of Scientific Biography,* s.v. "Fibonacci, Leonardo."
42. Haskins, *Renaissance,* 278–302. Menocal, *Arabic Role,* 9–10, 28–29, 48. Tronzo, *Intellectual Life at the Court of Frederick II.* Hassan and Hill, *Islamic Technology,* 32. Sarton, *From Rabbi Ben Ezra to Roger Bacon,* 491–495, 716–724. Long, *Technology and Society.*
43. White, "Tibet, India, Malaya," 520. Taccola, *De ingeneis* 2:80v, 3:30v in Prager and Scaglia, *Mariano Taccola.* Stuard, "Ancillary Evidence." Prunai, "Notizie e documenti sulla servitù," table follows page 438.
44. White, "Technology and Invention in the Middle Ages," 149. For the church's role in continuing certain urban administrative functions, see Squatriti, "Water and Society," and Ward-Perkins, *From Classical Antiquity to the Middle Ages.* A study that addresses the ways in which Cistercian

monasteries modified and exploited watercourses near Siena is Cortese, *L'acqua, il grano, il ferro.* For scholarship on Cistercians and technology elsewhere in Europe, see Benoit, "L'industrie cistercienne," as well as some of the other essays from Hetzlen, *Monachisme et technologie.* Hodge, *Roman Aqueducts and Water Supply.* Crouch, *Water Management.* Hughes, *Pan's Travail.* Smith, *Man and Water.*

45. Heurgon, *Daily Life of the Etruscans,* 105–6. Forbes, "Water Supply," 175–176. Glick, *Islamic and Christian Spain,* 226–227, 238.
46. *Dictionnaire des antiquités grecques et romains,* s.v. "Cloaca Maxima." The Etruscan influence upon Roman civilization remains a topic of hotly contested debate among archaeologists. Bonnin, "Hydrauliciens étrusques." "La campagne romaine, si fertile et si rich aux temps étrusques, n'a été, depuis l'Empire romain jusqu'aux cinquante dernières années, qu'une triste région désolée et infestée de malaria." Goblot, *Qanats,* 188–191. Heurgon, *Daily Life of the Etruscans,* 106.
47. Goblot credits Wertime's "Metallurgical Exploration," for placing the origins of mining in the Western ecumene somewhere on the Iranian plateau. *Qanats,* 60, 188–192. Hodge suggests an equally plausible route of diffusion of qanat-building skills during the Empire's expansion into areas where qanats already existed, such as the Levant and North Africa. *Roman Aqueducts,* 20.
48. Feruglio, "Approvvigionamento idrico nell'antica Perugia," 217.
49. Hodge, *Roman Aqueducts,* 346. Gardner, *Story of Siena,* 1. Hook, *Siena,* 9,10.
50. Dott.[sa] Marinella Vaglio, a Sienese resident of Colle di Val d'Elsa, relates that an architecture student at Florence wrote her thesis on the system in Colle in the 1990s. Unfortunately, such theses are not generally available to researchers. Conversation with author, July 1996. Oral report, meeting of La Diana, the Sienese aqueduct preservation organization, Fall 1993.
51. Riccetti, *Città costruita,* 160, 250. Waley, *Italian City-Republics,* 111, uses the figure "some two hundred feet."
52. *Acquedotto 2000,* 7, 42.
53. From the consortium of Exeter Museums' web page, "The History of Exeter's Underground Passages," at http://www.exeter.ac.uk/~rerandal/ALBERT/up-history.html, accessed 20 May 1998. I am indebted to Nicholas J. Wyatt of the Science Museum Library in London for bringing the Exeter system and the relevant literature to my attention. The one monograph, published by the Exeter City Museums and Art Gallery, is *Exeter's Underground Passages.*
54. The original is at Trinity College, Cambridge, Ms R.17.1, f.284v/285r. The map has been published in several places. See Greene's discussion of the diagram and the system at Canterbury in *Medieval Monasteries,* 109–111 and fig. 48 on p. 110. The best reproduction is in Grewe, *Wasserversorgung im Mittelalter,* 231–232, and a facsimile in the pocket inside the back cover. The role of the church is discussed in Ward-Perkins, *From Classical Antiquity to*

the Middle Ages; Squatriti, "Water and Society;" and Magnusson, *Water Technology in the Middle Ages.*

55. Ron, "Qanats and Spring Flow Tunnels," 212; idem, "Agricultural Terraces in the Judean Mountains;" idem, "Development and Management of Irrigation Systems."
56. Ron, "Qanats and Spring Flow Tunnels," 231.
57. Perhaps because their fieldwork has taken them to so many arid regions, the authors contributing to Beaumont, *Qanat, Kariz, and Khattara* and other students of qanats, such as Goblot, clearly understand the policy implications of their historical studies. Modern water supply engineering in Siena, which incorporates water from the medieval aqueducts into the citywide plan as a matter of course, shows that the industrialized world could also benefit from an understanding of ancient systems.

NOTES TO CHAPTER SIX

1. Bowsky, *Medieval Italian Commune,* 307.
2. White, "Flavor," 40. Scaglia, *Checklist,* 16. Keller, *Theatre of Machines.* Braunfels, *Mittelalterliche Stadtbaukunst.* Burckhardt, *Siena.* Bowsky, *Finance.* Hook, *Siena.* Goldthwaite, Building .
3. Braunfels, *Mittelalterliche Stadtbaukunst,* 121, 250–252; idem, *Urban Design,* 61–67. Lewis Mumford seems to be the first person to disseminate Braunfels's observations about Tuscan cities in English. See his *City in History,* 352, 585. Mumford, however, saw this widespread participation as an "internal weakness" because of the demands it made on each citizen's time.
4. Bowsky, *Nine,* 260, 298. Siena was not unique in its rewards system. Florence offered similar rewards as early as 1300 when it granted Arnolfo di Cambio tax exemption for life for his role as chief architect of the cathedral. Frisch, *Gothic Art,* 74.
5. Ettlinger, "Emergence of the Italian Architect."
6. Ackerman, "Architectural Practice," 4.
7. Ackerman, "Term 'Architect' in the Middle Ages."
8. Archivio dell'Opera del Duomo di Siena, Pergamena N. 233, published in Milanesi, *Documenti* 1:139. Archivio dell'Opera del Duomo di Siena, Pergamena N. 246, published in Milanesi, *Documenti,* 1:141. *Libri dell'entrata e dell uscita del comune di Siena detta della Biccherna: reg 29((1259 primo semestre),* 25, 39, 47, 61, 75, 126. Prager and Scaglia, *Mariano Taccola.* Chironi, "Repertorio."
9. Prager and Scaglia, *Mariano Taccola,* 3, 5, 8–9, 14–15, 17, 18, citing Milanesi, *Documenti,* 2:284–286; and Bacci, *Francesco di Valdambrino,* 367, ff. Enzo Carli believes he has identified four of these carvings. See his "Sculture del coro." Taccola's *denunzia* (a sort of tax declaration) of 1453 reads, in part, "During the last twenty years I had two occupations as officer

of the Commune of Siena, one as *stimatore* and the other as *viaio.*" For the various roles played by a *stimatore,* see Adams, "Life and Times." The attendants of the Casa di Misericordia e Sapientia were members of the Tertiary Order of the Humiliates, a semi-monastic charitable order.

10. Prager and Scaglia, *Mariano Taccola,* 222–226. All figures for numbers of sheets are derived from their index of published and unpublished editions of Taccola's manuscripts. Since each page can contain multiple versions of the same device, these numbers represent a minimum count.
11. Ferguson, "Mind's Eye."
12. Al-Muqaddasi, *Description de l'Afrique,* 124–5, cited in Hill, *History of Engineering,* 166. Hill, a civil engineer with long experience in the eastern Mediterranean, contends, "this application is of course, not possible in the Mediterranean." Prager and Scaglia, citing Dales, say that the first western description of the action of tides within the mouths of rivers was that of Robert Grosseteste, *Mariano Taccola,* 54, 135. Dales, "Grosseteste's *Questio de fluxu,*" 472. Prager and Scaglia argue that whereas tides were large enough to be economically useful in the Adriatic, they were not in the Tyrrhenian Sea. Squatriti, "Advent and Conquests," 130, 135–137.
13. Prager and Scaglia, *Mariano Taccola,* 226.
14. "Apparently he was a recorder of facts rather than a man of action or a philosopher. He was mainly an artist and a teacher." Prager and Scaglia, *Mariano Taccola,* 8, 21. Gille, *Engineers of the Renaissance,* 100. Scaglia, *Checklist,* 10, 18.
15. Much of the information in this paragraph is from Weller, *Francesco di Giorgio,* 2ff. Weller should be read in conjunction with Scaglia, *Checklist.* See also the recent exhibition catalog, Bellosi, *Francesco di Giorgio.* Chironi's review of the documentary evidence leaves most of Weller's findings substantially unchanged. Chironi uses Giorgio's *denunzia* (tax declaration) from the *Lira* (tax rolls or census) of the above years. Chironi, "Repertorio," 471–472, 482.
16. Consiglio Generale 232, Deliberazioni, 28 April 1469, ff. 274r-v, published in Weller, *Francesco di Giorgio,* which includes the text of the relevant documents in an appendix beginning on page 339. The payments are recorded in Biccherna. B. 328, f129v, 1469; B. 329, f57v, 1471; Biccherna 330, ff78r, 110r, 1472, and are published in Chironi, "Repertorio," 7, 472–473. Fiore, "Francesco di Giorgio," 757.
17. Scaglia, *Checklist,* 14.
18. Scaglia, *Checklist,* 10.
19. Scaglia, *Checklist,* 14, 25. Fiore, "Francesco di Giorgio," 753. Michelini, "Disegni e appunti."
20. Galluzzi, *Renaissance Engineers,* 35. Fiore, "Francesco di Giorgio," 753.
21. Reti, "Francesco di Giorgio," 287, 291.

22. Chironi, "Repertorio." Tucci, "Biringuccio, Vannoccio." Chironi, "Politici e ingegneri," 378–380. Prager and Scaglia, *Mariano Taccola,* 124–125, have an extended discussion of Taccola's role in transmitting military mines. They argue that first, Taccola did not invent the use of fire in mines (which can be traced back to Vegetius) or the use of gunpowder in mines (which they say Taccola apparently learned from Brunelleschi, who knew that Florence used it against Pisa in 1403). Nonetheless, Taccola's descriptions of the methods on ff. 47v, 48r, and 48v in his 1449 treatise *De machinis* appear to be the first visual depictions of it. Taccola, *De machinis,* 1:116–120, 2:79. Biringuccio wrote, "The first inventor of mines in Italy was surely Francesco di Giorgio Giorgi [i.e. Martini], an engineer and very excellent Sienese architect." Biringuccio alludes to the confusion already surrounding the origins of the technique by discussing the commonly held opinion that Pietro Navarra invented it. Biringuccio, *Pirotechnia,* 422–425.
23. Hauser, *Social History of Art,* 2:52–53.
24. Ettlinger, "Emergence of the Italian Architect," 96–123. Both Sienese and foreigners were able to work as engineers in Siena. In fact, it appears that foreigners were allowed to hold quite important posts; take, for example, the case when Giovanni di Niccolò da Pisa served as *capo maestro del Duomo* from 1285 to 1297. Hook, citing practical, financial, and aesthetic reasons, asserts that "other considerations being equal, Sienese citizens were chosen before other artists," *Siena,* 60.
25. Scaglia, *Checklist,* 13–14. For partnerships, see Chironi, "Politici e ingegneri," 375–395; idem, "Cultura tecnica e gruppo dirigente: la famiglia Vannocci Biringucci. Prime note." Siena, 1993. [Typescript courtesy of author.] Although Chironi drops the final "o," I follow the spelling of the *Dizionario biografico degli Italiani* and the spelling used in the English edition of the *Pirotechnia.* Frisch, *Gothic Art,* 51–90, provides examples from England, France, Florence, Siena and other places. Goldthwaite, *Building,* 352–356; idem, *Wealth and the Demand,* 178ff.
26. Reti, "Francesco di Giorgio." Ferguson, *Engineering and the Mind's Eye,* 70–74. Artz, *Development of Technical Education,* 30–33. White, "Flavor," 36-57.
27. Two studies that address this period are Ward-Perkins, *Classical Antiquity to the Middle Ages* and Squatriti, *Water and Society.* Hook, *Siena,* 57.
28. Burckhardt, *Siena,* 38. Ackerman, "'Ars Sine Scientia Nihil Est,'" 215, 222. Ackerman finds the use of the term *ingegnere* in Milanese documents in 1388 and 1389 (258).
29. The expansion for the grandiose New Duomo, which would have been the largest in Western Europe, began in 1316. A combination of structural and financial problems, compounded by the plague of 1348, resulted in the suspension of construction. Funds were nevertheless still being allocated to the

project as late as 1405. Bortolotti, *Siena,* 38–41. Bowsky, "Black Death." Hook, *Siena,* 97, 99, 103, 158.

30. Bowsky, "Black Death," 14, 19ff. Chiappelli, "Gli ordinamenti sanitari." "E in quella non valendo alcuno senno né humano provedimento, per lo quale fu da molte immondizie purgata la città da oficiali sopra ciò ordinati e vietato l'entrarvi dentro a ciascuno infermo e molti consigli data a coservazion della sanità. . . ." Boccaccio *Decameron,* "Introduzione," sentence 6., Branca ed., 1:14.
31. For the significance of improved methods of representing and transmitting ideas, see Ferguson, "Mind's Eye," 828; and idem, *Engineering and the Mind's Eye.*
32. Ackerman, "Architectural Practice," 4. Galluzzi, *Renaissance Engineers,* 27–28. Galluzzi discusses the lack of protection for intellectual property for engineer-inventors.
33. For a catalog illustrating the diffusion of Italian military engineers across Europe in the late-medieval and early-modern periods, see Maggiorotti, *Architetti.* Viganò addresses recent scholarship in "Architetti e ingegneri militari." 11–28. Galluzzi, "Macchine senesi," 23–25. Galluzzi, building upon Scaglia's 1971 suggestions, speculates that the humanist and ambassador, Ciriaco d'Ancona, could have presented the copy on one of his diplomatic missions to the Sultan. Ferguson, "Mind's Eye," 828.
34. White, "Flavor," 40-41.
35. Prager, "Manuscript of Taccola." See also Prager and Scaglia, *Mariano Taccola,* 11–12.
36. White, "Flavor," 50. Weller, *Francesco di Giorgio,* 18.
37. Judith Hook's chapter, "The Campo," provides a synthetic history of the Campo's evolution and an appreciation of its continuing role in Sienese life. *Siena,* 71–100.
38. Some examples include work in the Duomo by Nicola and Giovanni Pisano, the construction of the Torre di Mangia by Minuccio e Francesco di Rinaldo; work on the bottino maestro by three generations of the Jacopo della Acqua family; see also Vasari, *Lives,* especially the lives of Giotto, Pisano father and son, and Francesco di Giorgio. Kostof, "Architect in the Middle Ages." Hook, *Siena,* 29, 102.
39. Statuto 30 (1354–1392), f.16. Schultz and McShane describe the rediscovery of the advantages of tenuring urban engineers in the United States in "To Engineer the Metropolis." Milanesi, *Documenti,* 3:410, 412–413.
40. Bargagli-Petrucci, "Francesco di Giorgio." Civai, "Francesco di Giorgio, 153. Weller, *Francesco di Giorgio.* Bellosi, *Francesco di Giorgio.* For Taccola, see Prager and Scaglia, *Mariano Taccola,* 16. The reluctance to let one's competitors know the details of a proposal is illustrated in the story of "Brunelleschi's Egg," Vasari, *Lives,* 1:337. The reader can compare the accounts of the bronze competition by Lorenzo Ghiberti and his arch rival, Brunelleschi, in

Holt, *Documentary History of Art,* 1:157–158, 173–176. Ghiberti also collaborated with Jacopo della Quercia and Donatello on the sculptural program for Siena's baptismal font. Milanesi, *Documenti,* quoted and translated in Burckhardt, *Siena,* 37–39.

41. Gille, *Engineers of the Renaissance,* 9.

NOTES TO CHAPTER SEVEN

1. White, "Study of Medieval Technology."
2. Hoffmann, "Economic Development and Aquatic Ecosystems."
3. Drancourt, "Detection of 400-year-old *Yersinia pestis.*"

Bibliography

MANUSCRIPTS

(Located in the Archivio di Stato di Siena, unless otherwise noted.)

Bardi, Agnolo. "La seconda parte delle Istorie Senese." [1512–1556] ASS ms. D. 50.

Biccherna 35

Consiglio Generale

Concistoro, Deliberazioni

Fondo Ospedale di S. Maria della Croce, Archivio Comunale di Montalcino, Diplomatico, p. 133, n(472.

Ms D 50

Regolatori

Statuto 2

Statuto 5

Statuto 7

Statuto 17

Statuto 20

Statuto 38

Statuto dei Viarî 1

Statuto dei Viarî 2

Copy of Mariano Taccola. *De machinis.* Spencer ms 136. Division of Prints and Photographs, New York Public Library.

PUBLISHED PRIMARY SOURCES

Agricola, Georgius. *De Re Metallica.* Translated and edited by Herbert Clark Hoover and Lou Henry Hoover. 2d ed. New York, 1950.

Alberti, Leon Battista. *On the Art of Building in Ten Books.* Translated by Joseph Rykwert, Neil Leach, and Robert Tavernor. Cambridge, Mass., 1988.

Amari, Michele. *Biblioteca arabo-sicula: ossia raccolta di testi arabici che toccano la geografia, la storia, le biografie, e la bibliografia della Sicilia.* 2d ed. Edited by Umberto Rizzitano. Palermo, 1987.

Anonymous of the Hussite Wars. *The Technological Illustrations of the so-called "Anonymous of the Hussite Wars."* Edited and translated by Bert S. Hall. Wiesbaden, 1979.

Ascheri, Mario, ed. *L'ultimo statuto della Repubblica di Siena (1545).* Siena, 1993.

Astuti, Guido. *Il libro dell' entrata e dell' uscita di una compagnia mercantile senese del secolo XIII (1277-1282).* Turin, 1934.

Banchi, Luciano, and Filippo Luigi Polidori, eds. *Statuti Senesi scritti in volgare ne' secoli XIII e XIV e pubblicati secondo I testi del R. Archivio di Stato in Siena.* 3 vols. Bologna, 1863–77. Vol. 1 (1863): *Statuto del Comune di Montagutolo; Statuto dell'arte di Carnajuoli di Siena; Statuto dell'arte di Lana di Siena* (1298–1309); vol. 2 (1871): *Statuto della Gabella di Siena; Società del Padule d'Orgia; Arte della Lana di Radicondoli; Arte di Chiavari; Arte di Cuoiai e calzolai;* vol. 3 (1877): *Statuto dello spedale di Siena.*

Bargagli-Petrucci, Fabio. *Le fonti di Siena e i loro acquedotti: note storiche dalle origine fino al MDLV.* Siena, 1906.

Battini, Domenico. *Ricerche intorno alle acque minerali epatiche ed alla analisi chimica di diverse acque minerale dello stato di Siena.* Siena, 1799.

———. "Relazione intorno alle acque delle fonti di Siena: dei loro acquidotti, della loro natura ed usi, degli inconvenienti che ne derivano in pregiudizio della publica salute, e della maniera di ripararvi." *Atti dell' Accademia delle Scienze di Siena detta de' Fisiocritici* 8 (1800): 109–200. [a reprint of the above *Ricerche.*]

Biringuccio, Vannoccio. *The Pirotechnia of Vannoccio Biringuccio: The Classic Sixteenth Century Treatise on Metals and Metallurgy.* Translated and edited by Cyril Stanley Smith and Martha Teach Gnudi. New York, 1959. Reprint. New York, 1990.

Boccaccio, Giovanni. *Decameron.* Edited by Vittore Branca. 2 vols. Milan, 1980.

Boccaccio, Giovanni. *Decameron.* Translated by Mark Musa and Peter Bondanella. New York, 1982.

Bruni, Leonardo. "Panegyric to the City of Florence." In *The Earthly Republic: Italian Humanists on Government and Society.* Edited by Benjamin G. Kohl and Ronald G. Witt. Philadelphia, 1978.

Carducci, Giosue, Vittorio Fiorini, and Pietro Fedele, eds. *Raccolta degli storici italiani dal cinquecento al millecinquecento ordinata da L. A. Muratori.* Vol. 15, pt. 6, *Cronache Senesi.* Edited by Alessandro Lisini and Fabio Iacometti. Bologna, 1931–1939.

Catoni, Giuliano, ed. *Libri dell'entrata e dell'uscita del comune di Siena, detti della Biccherna: Reg. 30(1259 secondo semestre.* Rome, 1970.

Cecchini, Giovanni, ed. *Il Caleffo Vecchio del comune di Siena.* 3 vols. Siena, 1931-1940.

Cecchini, Giovanni, and Giulio Prunai, eds. *Brevi degli speziali: 1356–1542.* Vol. 1 of *Statuti volgare senesi.* Siena, 1942.

Chiappelli, Alberto, ed. "Gli ordinamenti sanitari del Comune di Pistoia contro la pestilenza del 1348." *Archivio Storico Italiano* ser. 4, no. 20 (1887): 3–24.

Chironi, Giuseppe, "Documenti inediti su Vannoccio Biringucci." In *Una tradizione senese dalla Pirotechnia di Vannoccio Biringucci al Museo del Mercurio.* Edited by Ivan Tognarini. Siena, 1992.

Ciampoli, Donatella. "Una raccolta di provvisioni senesi agli albori del XV secolo: il 'libro della catena,'" in *Bullettino senese di storia patria* 86 (1979): 243–283.

Ciampoli, Donatella, and Thomas Szabó, eds. *Viabilità e legislazione di uno stato cittadino del duecento: "Lo Statuto dei Viarî di Siena."* Siena, 1992.

de' Colli, Sandro. *Libri dell'entata e dell'uscita del comune di Siena. Detti della Biccherna.* Reg. 28 (1258 secondo semestre). Rome, 1965

Comini, Federico. *Rapporto alla Giunta Comunale sugli acquedotti che conducono le acque in Siena dell'assessore per gli affari legali.* Siena, 1873.

Cronaca Senese attributa ad Agnolo di Tura del Grasso, detta la Cronaca maggiore. In *Cronache Senesi.* Vol. 15, pt. 6 of *Rerum Italicarum Scriptores,* n.s. Edited by Alessandro Lisini and F. Iacometti. Bologna, 1931–1937.

Cronaca senese dall'anno 1202 al 1362 con aggiunte posteriori fina al 1391 di autore anonimo della metà del secolo XIV, 41–162. In *Cronache Senesi.* Vol. 15, pt. 6 *Rerum italicarum scriptores,* n.s. Edited by Alessandro Lisini and F. Iacometti. Bologna, 1931–1937.

Dales, Richard C., and Robert Grosseteste. "The Text of Robert Grosseteste's Questio De Fluxu Et Refluxu Maris with an English Translation." *Isis* 57 (1966): 455–74.

Dawson, Christopher, ed. *The Mongol Mission: Narratives and Letters of the Franciscan Missionaries in Mongolia and China in the Thirteenth and Fourteenth Centuries.* Translated by a Nun of Stanbrook Abbey. London, 1955.

Direzione del R. Archivio di Stato di Siena. *Libri dell'entrata e dell'uscita della Repubblica di Siena detti del camarlingo e dei quattro provveditori della Biccherna: Libri Primo e Secondo.* Siena, 1914. [Covers the years 1226–1229].

Direzione del R. Archivio di Stato di Siena. *Libri dell'entrata e dell'uscita della Repubblica di Siena detti del camarlingo e dei quattro provveditori della Biccherna: Quattordicesimo libro a.1253.* Siena, 1937.

Dondi, Giovanni de. *The Planetarium of Giovanni De Dondi, Citizen of Padua: A Manuscript of 1397.* Translated by G. H. Baillie, and H. Alan Lloyd. Edited by F. A. B. Ward. London, 1974.

Fantastici, Bernardino. *Campione di tutte le fabbriche, strade, piazze, fonti, acquidotti, canali e cloache pubbliche appartenenti alla comunità di Siena, 1789.* Edited by Carlo Cresti. Siena, 1992.

Fineschi, Sonia Adorni, ed. *Libri dell'entrata e dell'uscita del comune di Siena, detti della Biccherna: Reg. 29(1259 primo semestre.* Rome, 1969.

Fontana, Giovanni. *Le Macchine Cifrate di Giovanni Fontana: con la riproduzione del Cod. Icon. 242 della Bayerische Staatsbibliothek di Monaco di Baviera e la decrittazione di esso e del Cod. Lat. Nouv. Acq. 635 della Bibliothèque Nationale di Parigi.* Ed. Eugenio Battisti and Giuseppa Saccaro Battisti Milan, 1984.

Frontinus, Sextus Julius. *The Stratagems and the Aqueducts of Rome.* Translated by Mary B McElwain. Loeb Classical Library. London, 1925.

———. *De Aquaeductu Urbis Romae.* Translated by Robert H. Rodgers. Burlington, Vt. 2003. http://www.uvm.edu/~rrodgers/Frontinus.html (Accessed 7 May 2004).

Ghiberti, Lorenzo. *The Commentaries.* In *The Middle Ages and the Renaissance.* Vol. 1 of *A Documentary History of Art,* edited by Elizabeth G. Holt. New York, 1957.

al-Idrisi. *Il libro di Ruggero.* Translated and edited by Umberto Rizzitano. Palermo, 1966.

———. *L'Italia descritta nel "Libro del re Ruggero" comp. da Edrisi.* Translated and edited by Michele Amari and Celestino Schiaparelli. Rome, 1883.

al-Jazari. *The Book of Knowledge of Ingenious Mechanical Devices.* Edited and translated by Donald R. Hill. Dordrecht, 1974.

al-Karagi, Muhammad ibn al-Husayn. *La civilisation des eaux cachées: Traité de l'exploitation des eaux souterraines composé en 1017 A.D.* Translated and edited by Aly Mazaheri. Nice, 1973.

Lisini, Alessandro, ed. *Provvedimenti economici della repubblica di Siena nel 1382.* Siena, 1895.

Lisini, Alessandro, and Lodovico Zdekauer, eds. *Libri dell'entrata e dell'uscita della Repubblica di Siena detti del Camarlingo e dei Quattro Provveditori della Biccherna: Fasc. primo—Libro dell'anno 1226.* Siena, 1903.

Malavolti, Orlando. *Dell'historia di Siena.* 2 vols. in one. Venice, 1599.

Martini, Francesco di Giorgio. *Trattati di architettura ingegneria e arte militare.* Edited by Corrado Maltese. Milan, 1967.

Melis, Federigo, ed. *Documenti per la storia economica dei secoli XIII-XVI, con una nota di paleografica commerciale a cura di Elena Cecchi.* Florence, 1972.

Milanesi, Gaetano, ed. *Documenti per la storia dell'arte Senese.* 3 vols. Siena, 1854–1856.

Montaigne, Michel de. *The Journal of Montaigne's Travels in Italy by way of Switzerland and Germany in 1580 and 1581.* Translated and edited by W. G. Waters. 3 vols. London, 1903. 2:61.

Morandi, Ubaldo, ed. *Libri dell'entrata e dell'uscita del comune di Siena detta della Biccherna: Reg 27((1258 primo semestre).* Rome 1963.

Muratori, Lodovico Antonio. *Rerum italicarum scriptores. . . .* 25 vols. in 28. Milan, 1723-51.

Opsomer-Halleux, Carmélia, ed. *L'art de vivre en santé: Images et recettes du Moyen Âge: Le "Tacuinum Sanitatis," manuscrit 1041 de la Bibliothèque de l'Université de Liège.* Alleur, Belgium, 1991.

Pegolotti, Francesco Balducci. *La Practica della Mercatura.* Edited by Allan Evans. Cambridge, Mass., 1936.

Piattoli, Renato. *Lo statuto dell'arte dei padroni dei mulini sulla destra del fiume Bisenzio, 1296.* Prato, 1936.

Polo, Marco. *The Book of Ser Marco Polo, the Venetian: Concerning the Kingdoms and Marvels of the East.* Translated and edited by Colonel Sir Henry Yule. London, 1903.

Prunai, Giulio, "Notizie e documenti sulla servitù domestica nel territorio senese (sec. VIII - XVI)." *Bullettino senese di storia patria* 43 (1936): 133–182; 245–298; 398–438.

R. Archivio di Stato di Siena. *Il costituto del comune di Siena volgarizzato nel MCC-CIX-MCCCX.* Edited by Alessandro Lisini. 2 vols. Siena, 1903.

Santi, Giorgio. *Viaggio Terzo per le due Provincie Senesi che forma il seguito del Viaggio al Montamiata. . . .* Pisa, 1806.

Scalamonti, Francesco. *Vita viri clarisssimi et famosissimi Kyriaci Anconitani.* Translated and edited by Charles Mitchell and Edward W. Bodnar, S.J. Published as *Transactions of the American Philosophical Society* 86 (1996): 1–246.

Taccola, Mariano. *De Ingeneis: liber primus leonis, liber secundus draconis, addenda.* Edited by Gustina Scaglia, Frank D. Prager, and Ulrich Montag. Wiesbaden, 1984.

———. *De Machinis. The Engineering Treatise of 1449.* [Codex Latinus Monacensis 28800 in the Bayerishe Staatsbibliothek, München.] Translated and edited by Gustina Scaglia. Wiesbaden, 1971.

———. *De Machinis: De rebus militarius (De machinis, 1449).* Edited by Eberhard Knobloch. Baden-Baden, 1984.

———. *Liber tertius de ingeneis ac edifitiis non usitatis.* Edited by James H. Beck. Milan, 1969.

Tabula Peutingeriana: Codex Vindobonensis 324. Edited by Ekkehard Weber. Graz, 1976.

Tommasi, Giugurta. *Dell'historie di Siena.* 2 vols. in one. Venice, 1625–1626. Reprint. Bologna, 1973.

Venerosi-Pesciolini, Giulio. "*Ordinamenta balneorum* (MCCLXXXXII)" 129–135 of "I bagni senesi di Petriolo nel medioevo." *La Diana* 6 (1931): 110–135.

Vigo, Pietro. *Carlo Quinto in Siena nell'Aprile del 1536, relazione di un contemporaneo.* 1884. Reprint. Bologna, 1968.

Villani, Giovanni. *Selections from the First Nine Books of the Croniche Fiorentine of Giovanni Villani.* Translated by Rose E. Selfe. Edited by Philip H. Wicksteed. Westminster, 1896.

Vitruvius. *On Architecture.* 2 vols. Translated by Frank Stephen Granger. New York, 1931–1934.

Yule, Henry, ed. *Cathay and the Way Thither: Being a Collection of Medieval Notices of China.* 2 vols. London, 1866.

Zdekauer, Lodovico, ed. *Il Constituto del comune di Siena dell'anno 1262.* Milan, 1897. Reprint. Bologna, 1983.

SECONDARY SOURCES

Ackerman, James S. "The Term 'Architect' in the Middle Ages." *Speculum* 17 (1942): 549–562.

———. "Architectural Practice in the Italian Renaissance." *Journal of the Society of Architectural Historians* 13 (1954): 3-11.

———. "'Ars Sine Scientia Nihil Est': Gothic Theory of Architecture at the Cathedral of Milan." In *Distance Points: Essays in Theory and Renaissance Art and Architecture* Cambridge, Mass., 1991.

Acquedotto 2000: Bologna, l'acqua del duemila ha duemila anni. Casalecchio di Reno, Bologna, 1985.

Adams, Nicholas. "Architecture for Fish: The Sienese Dam on the Bruna River—Structures and Designs, 1468-ca. 1530." *Technology and Culture* 25 (1984): 768-97.

———. "The Life and Times of Pietro dell'Abaco, a Renaissance Estimator from Siena (active 1457–1486)." *Zeitschrift für Kunstgeschichte* 48 (1985): 384–395.

Adams, Nicholas, Daniela Lamberini, and Simon Pepper. "Un disegno di spionaggio cinquecentesco: Giovanni Battista Belluzzi e il rilievo delle difese di Siena ai Tempi dell'Assedio." *Mitteilungen des Kunsthistorischen Instituts in Florenz* 32 (1988): 558–79.

Allan, John, and Exeter City Museums and Art Gallery. *Exeter's Underground Passages.* Exeter, 1994.

Allen, Edward. *Stone Shelters.* Cambridge, Mass., 1969.

Artz, Frederick B. *The Development of Technical Education in France 1500–1800.* Cambridge, Mass., 1966.

Bacci, Peleo. *Francesco di Valdambrino, Emulo del Ghiberti e Collaboratore di Jacopo della Quercia.* Siena, 1936.

Baker, M. N. *The Quest for Pure Water: The History of Water Purification from the Earliest Records to the Twentieth Century.* 2d ed. 2 vols. N.p., 1981.

Balestracci, Duccio. "Approvvigionamento e distribuzione dei prodotti alimentari a Siena nell'epoca comunale: mulini, mercati, botteghe." *Archeologia Medievale* 8 (1981): 127–154.

———. "'Li lavoranti non cognosciuti.' Il salariato in una città medievale (Siena 1340–1344)." *Bullettino senese di storia patria* 82–83 (1975–76): 67–157.

———. "Val di Merse: profilo storico di un territorio tra Medioevo ed Età moderna." In *Sovicille,* edited by Roberto Guerrini. Milan, 1988.

———. "La Terra dei mulini." In *Monteroni: Arte, storia, territorio,* edited by Roberto Guerrini. [Sovicille, 1990].

———. "L'acqua a Siena nel Medioevo." In *Ars et Ratio: dalla Torre di Babele al ponte di Rialto,* edited by Jean-Claude Maire Vigueur and Agostino Paravicini Bagliani. Palermo, 1990.

———. "Regulation of Public Health in Italian Medieval Towns," in *Die Vielfalt der Dinge: neue wege zur Analyse mittelalterlicher Sachkultur. internationaler Kongre*

Krems an der Donau 4. Bis 7. Oktober 1994, edited by Helmut Hundsbichler, Gerhard Jaritz, and Thomas Kühtreiber. Vienna, 1998.

Balestracci, Duccio, and Gabriella Piccinni. *Siena nel Trecento: assetto urbano e strutture edilizie.* Florence, 1977.

Balestracci, Duccio, Daniela Lamberini, and Mauro Civai. *I bottini medievali di Siena.* Siena, [1993].

Banchi, Luciano. *I Porti della Maremma Senese durrante la Repubblica: narrazione storica con documenti inediti.* Florence, 1871.

———. "Provvisioni della Repubblica di Siena contro la Peste degli Anni 1411 e 1463." *Archivio Storico Italiano* Misc. LVI / 12 (1884).

Bargagli-Petrucci, Fabio, "Francesco di Giorgio Martini operaio dei bottini di Siena." *Bullettino senese di storia patria* 9 (1902): 227–236.

Barsini, G., and G. Bosco. "L'approvvigionamento idrico della città di Siena." *Nuovi Annali d'Igiene e Microbiologia* 18 (1966): 20–30.

Barsini, G., G. Bosco, and M. Prugnola. "Aspetti igienici e tecnici relativi all'allontamento dei rifiuti liquidi della città di Siena." *Nuovi Annali d'Igiene e Microbiologia* 17 (1966): 420–434.

Basanoff, Anne. *Itinerario della carta dall'Oriente all'Occidente e sua diffusione in Europa.* Milan, 1965.

Beals, Ralph. "Kroeber, Alfred L." In *International Encyclopedia of the Social Sciences.* 8:454–463

Beaumont, Peter, Michael Bonine, and Keith McLachlan, eds. *Qanat, Kariz, and Khattara: Traditional Water Systems in the Middle East and North Africa.* Wisbech, 1989.

Beck, James H. *Jacopo della Quercia.* 2 vols. New York, 1991.

———. "The Historical "Taccola" and Emperor Sigismund in Siena." *Art Bulletin* 50 (1968): 309–320. 1–30;

———. "Jacopo della Quercia and Donatello: Networking in the Quattrocento." *Source Notes in the History of Art* 6 (1987): 6–15.

Bedini, Silvio A., and Francis R. Maddison. *Mechanical Universe: The Astrarium of Giovanni de' Dondi.* Transactions of the American Philosophical Society, n.s., no. 56:2, part 5. Philadelphia, 1956.

Bellosi, L. ed. *Francesco di Giorgio e il Rinascimento a Siena 1450–1500.* Milan, 1993.

Benoit, Paul. "L'industrie cistercienne: les raisons d'une indifférence (XIIème siècle-1ère moitié du XIVème siècle)," In *Monachisme et technologie dans la société médiévale du Xe au XIIIe siècle,* edited by Charles Hetzlen and René de Vos. Paris, 1994.

Bentley, Jerry H. *Old World Encounters: Cross-Cultural Contacts and Exchanges in Pre-Modern Times.* New York, 1993.

Bergamini, Margherita. *Gli Etruschi: maestri di idraulica.* Perugia, 1991.

Bermejo, J. Esteban Hernández, et al. "Economic Botany and Ethnobotany in Al-Andalus (Iberian Peninsula: Tenth-Fifteenth Centuries), an Unknown Heritage of Mankind." *Economic Botany* 52 (1996): 15–26.

Bezzini, Mario. *Formazione e sviluppo di Siena medievale.* Siena, 1981.

Betti, Luca, and Alessandro Falassi. *Com'era Siena.* Siena, 1993.

Bizzarri, Dina. "Tentativi di bonifiche nel contado Sense nei secoli XIII-XIV." *Bullettino senese di storia patria* 24 (1917): 131–160.

Blake, Nelson Manfred. *Water for the Cities.* Syracuse, 1956.

De La Blanchère, R. "Cuniculus." In *Dictionnaire des antiquités grecques et romains,* edited by Charles Daremberg and Edmond Saglio. 5 Vols. Paris, 1877–1919.

Bloch, Marc. "Avènement et conquête du moulin à eau." *Annales d'histoire économique et sociale* 7 (1935): 538–53.

Boethius, Axel, and J. B. Ward-Perkins. *Etruscan and Early Roman Architecture.* Harmondsworth, 1970.

Boisseuil, Didier. "Les stations thermales dans le territoire siennois de la fin du XIIIe siècle au milieu du XVe siècle, quelques aspects économiques." *Tempo libero* (1995): 731–741.

———. *Le thermalisme en Toscane à la fin du Moyen Âge: les bains siennois de la fin du XIIIe siècle au début du XVIe siècle.* Rome, 2002.

———. "Les Bains siennois de la fin du XIIIe siècle au début du XVIe siècle. Contribution à l'histoire du thermalisme toscan au Moyen Age." Ph.D. diss., Université François Rabelais, Tours, 1996.

Bono, P., and C. Boni. "Water Supply of Rome in Antiquity and Today." *Environmental Geology* 27 (1996): 126–134.

Bonnin, Jacques. "Les hydrauliciens étrusques, des précurseurs." *La houille blanche* 8 (1973): 641–49.

———. *L'eau dans l'antiquité: L'hydraulique avant notre ere.* Paris, 1984.

Bortolotti, Lando. *Siena.* Rome, 1983.

———. "Studi recenti di storia urbana su Siena." *Storia Urbana* 30 (1985): 141–154.

Bowsky, William M. *The Finance of the Commune of Siena 1287-1355.* Oxford, 1970.

———. "The Constitution and Administration of a Tuscan Republic in the Middle Ages and Early Renaissance: The *Maggior Sindaco* in Siena," *Studi Senesi* 80 (1968): 7–22.

———. "Ferdinand Schevill and the History of Mediaeval Siena: A Critical Essay." Introduction to *Siena: The History of a Medieval Italian Commune,* by Ferdinand Schevill. New York, 1909. Reprint. New York, 1964.

———. "The Impact of the Black Death upon Sienese Government and Society," *Speculum* 39 (1964): 1–34.

———. *A Medieval Italian Commune: Siena under the Nine. 1287–1355.* Berkeley, 1981.

———. "The Sienese Archive and the Pubblicazioni degli archivi di stato." *Manuscripta* 5 (1961): 67-77.

———. "Siena." In *The Dictionary of the Middle Ages,* edited by Joseph L. Strayer. 13 vols. New York, 1982-1989. 11:277–282.

———. "The *Buon Governo* of Siena (1287–1355): A Mediaeval Italian Oligarchy." *Speculum* 37 (1962): 368–381.

Braudel, Fernand. *The Mediterranean and the Mediterranean World in the Age of Philip II.* 2 vols. Translated by Siân Reynolds. New York, 1972.

———. *The Structures of Everyday Life: The Limits of the Possible.* Vol. 1 of *Civilization and Capitalism 15th - 18th Century.* Translation from the French revised by Siân Reynolds. New York, 1985.

Braunfels, Wolfgang. *Mittelalterliche Stadtbaukunst in der Toskana.* 5th ed. Berlin, 1982.

———. *Urban Design in Western Europe: Regime and Architecture, 900–1900.* Translated by Kenneth J. Northcott. Chicago, 1988.

———. *Monasteries of Western Europe: The Architecture of the Orders.* Translated by Alastair Laing. Princeton, 1972.

Braunstein, Philippe. "Gli Statuti minerari nel Medioevo europeo." In *Archeologia delle attività estrattive e metallurgiche,* edited by Riccardo Francovich. Florence, 1993.

Brilli, Attilio. *Viaggiatori stranieri in terra di Siena.* Rome, 1986.

Brooks, Kenneth N., et. al. *Hydrology and the Management of Watersheds.* 2d ed. Ames, 1997.

Brown, Judith C. "Prosperity or Hard Times in Renaissance Italy?" *Renaissance Quarterly* 42 (1989): 761–780.

Bufalini, Baldassarre. "Igiene." In *Siena e il suo territorio,* edited by Luigi Lazzeri. Siena, 1862. Reprint. Bologna, 1978.

Buffet, Bernard, and René Evrard. *L'Eau Potable a Travers les Ages.* Liège, 1951.

Burckhardt, Titus. *Siena: City of the Virgin.* Translated by Margaret McDonough Brown. London, 1960.

Burns, Alfred. "Ancient Greek Water Supply and City Planning: A Study of Syracuse and Arcagas." *Technology and Culture* 15 (1974): 389–412.

Caferro, William. "City and Countryside in the Second Half of the Fourteenth Century." *Journal of Economic History* 54 (March 1994): 85–103.

———. *Mercenary Companies and the Decline of Siena.* Baltimore, 1998.

Cairola, Aldo, and Enzo Carli. *Il Palazzo Pubblico di Siena.* Siena, 1963.

Calchini, Enrica. "Glossario dei termini tecnici nel Trattato I (Ms. Saluzziano 148) di Francesco di Giorgio," in *Prima di Leonardo, cultura delle macchine a Siena nel Rinascimento,* edited by Paolo Galluzzi. Milan, 1991.

Cambi, Vincenzo. *Del nuovo uso delle acque di Fontebranda. Progetto proposto al Municipio di Siena.* Siena, 1869.

Camporeale, Giovannangelo, ed. *L'Etruria mineraria.* Milan, [1985].

Canestrelli Antonio. *L'Abbazia di San Galgano: monografia storico-artistica con documenti inediti e numerose illustrazioni.* Firenze, 1896.

Carli, Enzo. *Il Duomo di Siena.* Florence, 1966.

———. "Le sculture del coro del Duomo di Siena." *Antichità viva* 17:1 (1978): 25-39.

Carmichael, Ann G. "Plague Legislation in the Italian Renaissance." *Bulletin of the History of Medicine* 57 (1983): 508-525.

Cardini, Franco. "L'acqua come simbolo nel giardino toscano tardo medievale." In *L'Eau dans la société médiévale: Fonctions, enjeux, images,* edited by Jacques Dalarun. Rome, 1992. *MEFRM: MA* 104 (1992): 545–553.

Cardini, Franco, et al. *Banchieri e mercanti di Siena.* Preface by Carlo Cipolla. Rome, 1987.

Carus-Wilson, E. M. "An Industrial Revolution of the Thirteenth Century." *The Economic History Review* 11 (1941): 39–60.

———. "The Woolen Industry." In *Trade and Industry in the Middle Ages,* edited by M. M. Postan and Edward Miller. Cambridge, 1966.

Cassirer, Ernst, Paul Kristeller, and John Herman Randall, Jr., eds., *The Renaissance Philosophy of Man.* Chicago, 1948.

Casanova, Eugenio. "La donna senese nel Quattrocentro." *Bullettino senese di storia patria* 8 (1901): 3–93.

Caselli, Giovanni. *Guida alle antiche strade romane.* Revised by Renzo Rossi. Novara, 1994.

Caskey, Jill. "Steam and *Sanitas* in the Domestic Realm: Baths and Bathing in Southern Italy in the Middle Ages." *Journal of the Society of Architectural Historians* 58 (1999): 170–195.

Castiglioni, Arturo. "Ugo Benzi da Siena e il 'Trattato utilissimo circa la conservazione della sanitade.'" *Rivista di storia critica delle scienze mediche e naturali* 12 (1921): 75–102.

Cavallucci, Francesco. *La Fontana maggiore di Perugia: voci e suggestioni di una comunità medievale.* Ponte San Giovanni, Perugia, 1993.

Cecchini, Giovanni. "L'Archivio dell' Opera del Duomo e il suo riordinamento e inventario." *Bullettino senese di storia patria* 60 (1953): 67–77.

———. "Dove era la Siena romana?" *Il Campo di Siena* 6, no. 271 (4 Sept 1957), 3.

———. "Le grance dell'Ospedale di S. Maria della Scala di Siena." *Economia e storia* 6 (1959): 405–422.

Ceppari, Maria Assunta. "Tra legislazione annonaria e tecnologia: alla ricerca di una Biccherna perduta." In *Antica Legislazione della Repubblica di Siena,* edited by Mario Ascheri. Siena, 1993.

Cherubini, Giovanni. *Uomini, terre e città nel medioevo.* Vita civile degli italiani: società, economia, cultura materiale. Milan, 1986.

Chevallier, Raymond. *Roman Roads.* Translated by N. H. Field. Berkeley 1976.

Chiostri, Frido. *L'acquedotto romano di Firenze. Considerazioni generali e note tecnologiche sulla morfologia strutturale e sulle fasi operative della costruzione.* Florence, 1973.

Chironi, Giuseppe. "Politici e ingegneri: i provveditori della Camera del Comune di Siena negli anni '90 del Quattrocento." *Ricerche Storiche* 23 (1993): 375–395.

———. "Cultura tecnica e gruppo dirigente: la famiglia Vannocci Biringucci. Prime note." Siena, 1993. [Typescript courtesy of author.]

———. "Repertorio dei documenti riguardanti Mariani di Iacopo detto il Taccola e Francesco di Giorgio Martini." In *Prima di Leonardo: cultura delle macchine a Siena nel Rinascimento,* edited by Paolo Galluzzi. Milan, 1991.

———. "Appendice documentaria." In *Francesco di Giorgio e il Rinascimento a Siena 1450–1500,* edited by L. Bellosi. Milan, 1993.

Cipolla, Carlo M. *I pidocchi e il Granduca: crisi e problemi sanitari nella Firenze del '600.* Bologna, 1979.

Citarella, Armand O. "The Relations of Amalfi with the Arab World before the Crusades." *Speculum* 42 (1967): 299–312.

Civai, Mauro. "Francesco di Giorgio: operaio dei bottini di Siena," In *I bottini medievali di Siena,* edited by Balestracci, Duccio, Daniela Lamberini, and Mauro Civai. Siena, [1993].

Clagett, Marshall. "The Life and Works of Giovanni Fontana." *Annali dell'Istituto e Museo di Storia della Scienza di Firenze* 1 (1976): 5–28.

Comune di Siena and Duccio Balestracci. *I bottini: acquedotti medievali senesi.* Siena, 1984.

Constable, Olivia Remie. *Trade and Traders in Muslim Spain: The Commercial Realignment of the Iberian Peninsula.* Cambridge, 1994.

————, ed. *Medieval Iberia: Readings from Christian, Muslim, and Jewish Sources.* Philadelphia, 1997.

Costantini, B. "Sienne." In *Le Tour du Monde: Nouveau journal des voyages.* 11–12 (1866): 1–32.

Cortese, Maria Elena. *L'acqua, il grano, il ferro: opifici idraulici medievali nel bacino Farma-Merse.* Florence, 1997.

Coscarella, Giuseppina, and Franca Cecilia Franchi. *La Grancia di Cuna in Val d'Arbia: un esempio di fattoria fortificata medievale.* Florence, 1983.

Coutant, Yves. "L' étude des moulins à vent médiévaux: nouvelles pistes." *Revue du nord* 74 (1992): 5–24.

Cristofani, Mauro, ed. *Siena: le origini, testimonianze e miti archeologici.* Florence, 1979.

Crouch, Dora. "Modern Insights from the Study of Ancient Greek Water Management." In *Future Currents in Aqueduct Studies,* edited by A. Trevor Hodge. Leeds, 1991.

———. *Water Management in Ancient Greek Cities.* Oxford, 1993.

Dean, Trevor, and Chris J. Wickham, eds. *City and Countryside in Late Medieval and Renaissance Italy.* London, 1988.

De Solla Price, Derek J. "On the Origin of Clockwork, Perpetual Motion Devices and the Compass." *United States National Museum Bulletin 218: Contributions from the Museum of History and Technology Papers 1–11* (Washington, 1959): 81–112.

———. *Gears from the Greeks: The Antikythera Mechanism, A Calendar Computer from ca. 80 B.C.* New York, 1975.

De Vecchi, Vittoria. "L'architettura Gotica civile Senese." *Bullettino senese di storia patria* 56 (1949): 3–52.

De Roover, Raymond. "A Florentine Firm of Cloth Manufacturers: Management and Organization of a Sixteenth-Century Business." *Speculum* 16 (1941): 1–33.

Dini, Vittorio. *Il potere delle antiche madri fecondità e culti delle acque nella cultura subalterna toscana.* Turin, 1980.

Douglas, R. Langton. *History of Siena.* London, 1902.

Drancourt. Michel, et al. "Detection of 400-year-old *Yersinia pestis* DNA in human dental pulp: An approach to the diagnosis of ancient septicemia." *Proceeding of the National Academy of Sciences* 95 (October 1998): 12637–12640.

Edizione archeologica della carta d'Italia al 100.000, foglio 120 (Siena). Edited by Rannuccio Banchi Bandinelli. Florence, 1927.

Edler, Florence. *Glossary of Medieval Terms of Business: Italian Series 1200–1600.* Cambridge, Mass., 1934.

English, Edward D. *Enterprise and Liability in Sienese Banking, 1230 - 1350.* Cambridge, Mass., 1988.

English, P. W. "The Origin and Spread of Qanats in the Old World." *American Philosophical Society Proceedings* 112 (1968): 170–181.

Epstein, Stephan R. *Alle origini della fattoria toscana: l'Ospedale della Scala di Siena e le sue terre (metà '200 - metà '400.* Florence, 1986.

Ettlinger, Leopold D. "The Emergence of the Italian Architect during the Fifteenth Century," in *The Architect: Chapters in the History of the Profession,* edited by Spiro Kostof. New York, 1977.

Evans, Harry B. *Water Distribution in Ancient Rome: The Evidence of Frontinus.* Ann Arbor, 1997.

Exeter Excavations Committee. "Report on the Underground Passages of Exeter" in the *Proceedings of the Devon Archaeological Exploration Society* 1:4 (1932): 191-200.

Fabiano, Caterina Laganara. "L'acqua: i suoi contenitori nel medioevo. Alcuni esempi in Puglia." In *L'Eau dans la société médiévale: Fonctions, enjeux, images,* edited by Jacques Dalarun. Rome, 1992. *Mélanges de l'École Française de Rome: Moyens âges, temps modernes* 104 (1992): 373–386.

Falletti-Fossati, Carlo. *Costumi Senesi nella seconda metà del secolo XIV.* Siena, 1881. Reprint. Siena, 1980.

Feruglio, Anna Eugenia. "Approvvigionamento idrico nell'antica Perugia: la cisterna di via Cesare Caporali," in *Gli Etruschi: maestri di idraulica,* ed. by Margherita Bergamini Perugia, 1991.

Ferguson, Eugene S. "The Mind's Eye: Nonverbal Thought in Technology." *Science* 197 (1977): 827–836.

———. *Engineering and the Mind's Eye.* Cambridge, Mass., 1992.

Fiore, F. P. "Francesco di Giorgio." In *Dizionario biografico degli Italiani* 49: 753–759.

Forbes, R. J. "Water Supply." In *Studies in Ancient Technology.* 9 vols. to date. Leiden, 1955-. 1:145–184.

———. "Washing, Bleaching, and Felting." In *Ancient Technologies* Leiden, 1966. 4:81–89

———. "Leather in Antiquity." In *Ancient Technologies.* 2d ed., 9 vols to date. Leiden, 1966. 5:1–9.

———. "Hydraulic Engineering and Sanitation." In *The Mediterranean Civilizations and the Middle Ages, c. 700 B.C. to c. 1500 A.D.* Vol. 2 of *A History of Technology,* edited by Charles Singer, et al. Oxford, 1957.

Francovich, Riccardo. *Archeologia delle attività estrattive e metallugiche.* Florence, 1993.

———. *Carta archeologica della provincia di Siena.* Siena, 1995.

Frisch, Teresa. *Gothic Art 1140-c 1450: Sources and Documents.* 1971. Reprint. Toronto, 1987.

Gabbitas, Victoria. "Leather and Leatherworking." In *Dictionary of the Middle Ages,* edited by Joseph R. Strayer, 13 vols. New York, 1982–1989. 7:530–532.

Galluzzi, Paolo, ed. *Prima di Leonardo: cultura delle macchine a Siena nel Rinascimento.* Milan, 1991.

———. "Le macchine senesi: ricerca antiquaria, spirito di innovazione e cultura del territorio." In *Prima di Leonardo: cultura delle macchine a Siena nel Rinascimento.* Milan, 1991.

———. *Renaissance Engineers: From Brunelleschi to Leonardo da Vinci.* Florence, 1996.

Gams, Pius Bonifacius. *Series episcoporum ecclesiae catholicae.* 1873. Reprint. Graz, 1957.

Gardner, Edmund G. *The Story of Siena and San Gimignano.* London, 1904.

Garosi, Alcide. *Siena nella storia della medicina (1240–1555).* Florence, 1958.

Gensini, Sergio, ed. *La Toscana nel secolo XIV: caratteri di una civiltà regionale.* Pisa, 1988.

Gerschenkron, Alexander. *Economic Backwardness in Historical Perspective.* Cambridge, Mass., 1962.

Ghazanfar, S. M. "History of Economic Thought: The Schumpeterian 'Great Gap,' The 'Lost' Arab-Islamic Legacy and the Literature Gap." *Journal of Islamic Studies* 6 (1995): 234–253.

Gies, Frances, and Joseph Gies. *Cathedral, Forge, and Waterwheel: Technology and Invention in the Middle Ages.* New York, 1995.

Gille, Bertrand. *Engineers of the Renaissance.* Cambridge, Mass., 1966.

Gimpel, Jean. *The Medieval Machine.* New York, 1976.

Ginatempo, Maria. *Crisi di un territorio: il popolamento della Toscana senese alla fine del medioevo.* Biblioteca storica Toscana 24. Florence, 1988.

———. "Per la storia demografica del territorio senese nel Quattrocento: problemi di fonti e di metodo." *Archivio Storico Italiano* 422 (1984): 511–587.

———. "Per la storia degli ecosistemi e dell'alimentazione medievali: recenti studi di archeozoologia in Italia." *Archeologia Medievale* 11 (1984): 35–61.

Glick, Thomas F. *Irrigation and Society in Medieval Valencia.* Cambridge, Mass., 1970.

———. *Islamic and Christian Spain in the Early Middle Ages.* Princeton, 1979.

———. "Toledo." *Dictionary of the Middle Ages,* 12:67–68.

———. "Moriscos and Marranos as Agents of Technological Diffusion." *History of Technology* 17 (1995): 113–125.

Goblot, Henri. *Les qanats: Une technique d'acquisition de l'eau.* Paris, 1979.

Goitein, S. D. *A Mediterranean Society: The Jewish Communities of the Arab World as Portrayed in the Documents of the Cairo Geniza.* 6 vols. Berkeley, 1967–1993.

Goldthwaite, Richard A. *The Building of Renaissance Florence: An Economic and Social History.* Baltimore, 1980.

———. *Wealth and the Demand for Art in Italy, 1300–1500.* Baltimore, 1993.

Goldsmith, Edward. "Learning to Live with Nature." *The Ecologist* 28 (1998): 162–171.

Goudie, Andrew. *The Human Impact on the Natural Environment.* 5th ed. Cambridge, Mass., 2000.

Goss, Vladimir P., and Christine Verzár Bornstein, eds. *The Meeting of Two Worlds: Cultural Exchange between East and West During the Period of the Crusades.* Studies in Medieval Culture 21. Kalamazoo, 1986.

Grande Dizionario Della Lingua Italiana. Edited by Tullio De Mauro. 21 vols, Turin, 2002.

Greene, J. Patrick. *Medieval Monasteries.* Leicester, 1992.

Grewe, Klaus, et al. *Die Wasserversorgung im Mittelalter.* Vol. 4 of *Geschichte der Wasserversorgung.* Edited by the Frontinus Gesellschaft. Mainz am Rhein, [1991].

Guerrini, Silvano. *'Maria lavava': curandai e lavandai da Grassina a Rimaggio.* Grassina, 1991.

Guidoni, Enrico. *Il Campo di Siena.* Rome, 1971.

Guillerme, André E. *The Age of Water: The Urban Environment in the North of France, A.D. 300-1800.* College Station, Texas, 1988.

Hall, Bert S. "Lynn White's *Medieval Technology and Social Change* after Thirty Years." In *Technological Change: Methods and Themes in the History of Technology,* edited by Robert Fox. n.p. 1996.

———. "Giovanni de' Dondi and Guido da Vigevano: Notes Toward a Typology of Medieval Technological Writings." In *Machaut's World: Science and Art in the Fourteenth Century,* edited by Madeleine Pelner Cosman and Bruce Chandler. Annals of the New York Academy of Sciences 314. New York, 20 October 1978.

Hall, Peter Geoffrey. *Cities in Civilization: Culture, Innovation, and Urban Order.* London, 1998.

Hanson, Anne Coffin. *Jacopo della Quercia's Fonte Gaia.* Oxford, 1965.

Haskins, Charles Homer. *The Renaissance of the Twelfth Century.* Cambridge, Mass., 1927.

al-Hassan, Ahmad Y., and Donald R. Hill. *Islamic Technology: An Illustrated History.* Cambridge, 1986.

Hauser, Arnold. *Social History of Art.* Translated by Stanley Godman. London, 1951.

Herlihy, David. *The Black Death and the Transformation of the West.* Edited by Samuel K. Cohn, Jr. Cambridge, Mass., 1997.

Hersey, George. "Water-Works and Water-Play in Renaissance Naples." In *Fons Sapientiae: Renaissance Garden Fountains,* edited by Elisabeth MacDougall. Washington, 1978.

Charles Hetzlen and René de Vos, eds. *Monachisme et technologie dans la société médiévale du Xe au XIIIe siècle.* Paris, 1994.

Heurgon, Jacques. *Daily Life of the Etruscans.* Translated by James Kirkup. New York, 1964.

Heyd, Wilhelm von. *Histoire du commerce du Levant au moyen age.* 2d ed. Leipzig, 1923.

Hill, Donald R. *A History of Engineering in Classical and Medieval Times.* London, 1984.

———. "Information on Engineering in the Works of Muslim Geographers." *History of Technology* 9 (1984): 127–142.

———. *Islamic Science and Engineering.* Edinburgh, 1993.

Hills, Richard L. *Power from Wind: A History of Windmill Technology.* Cambridge, 1994.

Hodge, A. Trevor. *Roman Aqueducts and Water Supply.* London, 1992.

————, ed. *Future Currents in Aqueduct Studies.* Leeds, 1991.

Hoffmann, Richard C. "Economic Development and Aquatic Ecosystems in Medieval Europe." *American Historical Review* 101 (1996): 631–669.

Holmes, Urban Tigner. *Daily Living in the Twelfth Century: Based on the Observations of Alexander Neckam in London and Paris.* Madison, 1952.

Holt, Richard. "Mechanization in the Medieval English Economy." In *Technology and Resource Use in Medieval Europe: Cathedrals, Mills, and Mines,* edited by Elizabeth Bradford Smith and Michael Wolfe. Aldershot, 1997.

———. *The Mills of Medieval England.* Oxford, 1988.

Hook, Judith. *Siena: A City and Its History.* London, 1979.

———. "Water Power and the Plan of St. Gall." *Journal of Medieval History* 1 (1975): 219–58.

Hoshino, Hidetoshi. "The Rise of the Florentine Woollen Industry in the Fourteenth Century." In *Cloth and Clothing in Medieval Europe,* edited by N. B. Harte and K. G. Ponting. London, 1983.

Hughes, J. Donald. *Pan's Travail: Environmental Problems of the Ancient Greeks and Romans.* Baltimore, 1994.

L'Immagine di Siena: Le Due Città. Le piante degli acquedotti sotterranei di Siena nelle collezioni cittadine (dal XVI al XIX secolo). Exhibition catalog. Siena, 1999.

Irigoin, Jean. "Les Origines de la fabrication du papier en Italie." *Papiergeschicte* 13 (1963): 62–67.

Italy, Servizio Geologico. *Carta Geologica d"Italia al 1:500,000.* Sheet 2 and Text 2. Rome, 1978.

Jones, Philip. *The Italian City-State: From Commune to Signoria.* Oxford, 1997.

Judson, Sheldon, and Anne Kahane. "Underground Drainageways in Southern Etruria." *Papers of the British School in Rome* 31 (1963): 74–99.

Jung, J. "Das Itinerar des Erzbischofs Sigeric von Canterbury und die Strasse von Rom über Siena nach Lucca," *Mitteilungen des Instituts für Österreichische Geschichtsforschung* 25 (1904): 1–90.

Kealey, Edward J. *Harvesting the Air: Windmill Pioneers in Twelfth-Century England.* Berkeley, 1987.

Keller, Alex. *A Theatre of Machines.* New York, 1964.

Kempers, Bram. *Painting, Power and Patronage: The Rise of the Professional Artist in the Italian Renaissance.* Translated by Beverley Jackson. London, 1992.

Klaubert, Helmut. "Qanats in an Area of Bavaria-Bohemia." *The Geographical Review* 57 (1967): 203–212.

Konvitz, Josef. *The Urban Millennium: The City Building Process from the Early Middle Ages to the Present.* Carbondale, Ill., 1985.

Kostof, Spiro, "The Architect in the Middle Ages, East and West." In *The Architect: Chapters in the History of the Profession.* New York, 1977.

Kroeber, Alfred L. *Anthropology.* New York, 1923.

———. "Stimulus Diffusion." *American Anthropologist* 42 (1940): 1–20.

Krueger, Hilmar C. "The Italian Cities and the Arabs before 1095." In *The First Hundred Years,* edited by Marshall W. Baldwin. Vol. 1 of *A History of the Crusades,* 2d ed., edited by Kenneth M. Setton. Madison, 1969. Mass., 1983.

Langdon, John. "Review of *Harvesting the Air,*" *Speculum* 64 (1989): 451–453

Larner, John. *Culture and Society in Italy, 1290–1420.* London, 1971.

Lawrence, D. H. *Etruscan Places.* New York, 1932.

Lightfoot, Dale R. "Syrian Qanat Romani: History, Ecology, Abandonment." *Journal of Arid Environments* 33 (1996): 321–326.

———. "Qanats in the Levant: Hydraulic Technology at the Periphery of Early Empires." *Technology and Culture* 38 (1997): 432–451.

Lillich, Meredith Parsons. "Cleanliness with Godliness: A Discussion of Medieval Monastic Plumbing." In *Mélanges à la Mémoire du Père Anselme Dimier,* edited by Benoît Chauvin. Arbois, 1982.

Lindberg, David C. *The Beginnings of Western Science: The European Scientific Tradition in Philosophical, Religious and Institutional Context, 500 B.C. to A.D. 1450.* Chicago, 1992.

Lisini, Alessandro. "Notizie delle miniere della Maremma toscana e leggi per l'estrazione dei metalli nel Medievo." *Bullettino senese di storia patria* 42 (1935): 185–256.

Long, Pamela O. *Technology and Society in the Medieval Centuries: Byzantium, Islam, and the West, 500-1300.* Washington, 2003. .

———. "Invention, Authorship, 'Intellectual Property,' and the Origin of Patents: Notes toward a Conceptual History." *Technology and Culture* 32 (1991): 846–884.

———. "The Contribution of Architectural Writers to a 'Scientific' Outlook in the Fifteenth and Sixteenth Centuries." *Journal of Medieval and Renaissance Studies* 15 (1985): 265–298.

Lopez, Robert S. "European Merchants in the Medieval Indies: The Evidence of Commercial Documents," *Journal of Economic History* 3 (1942): 164–184.

———. "Silk Industry in the Byzantine Empire." *Speculum* 20 (1945): 1–42

———. "China Silk in Europe in the Yuan Period." *Journal of the American Oriental Society* 72 (1952): 72–76.

———. "L'importanza del mondo islamico nella vita economica europea." In *L'Occidente e L'Islam nell'Alto Medioevo,* 433–460. Spoleto, 1965.

———. "Hard Times and Investment in Culture." In *Social and Economic Foundations of the Italian Renaissance,* edited by Anthony Molho. New York, 1969.

Lopez, Robert S., and Irving W. Raymond, eds. and trans. *Medieval Trade in the Mediterranean World.* New York, 1955. Reprint. New York, 1965.

Luchini, Luca, and Piero Ligabue. *Siena dei bisnonni.* Siena, 1986.

Luchini, Luca. *Siena dei nonni.* Monteriggioni, 1993.

Lusini, Vittorio. "Note storiche sulla topografia di Siena nel sec XIII." *Bullettino senese di storia patria* 28 (1921): 239–341.

Maggiorotti, Leone Andrea. *Architetti e architetture militari.* Rome, 1933.

Magnusson, Roberta Jean. "Medieval Water Supplies: Hydraulic Technology and Social Organization in England and Italy." Ph.D. diss., University of California, Berkeley, 1994.

———. *Water Technology in the Middle Ages: Cities, Monasteries, and Waterworks after the Roman Empire.* Baltimore, 2001.

Magnusson, Roberta, and Paolo Squatriti. "The Technologies of Water in Medieval Italy." In *Working with Water in Medieval Europe: Technology and Resource Use.* Leiden, 2000.

Mariani, Maria Stella Calò. "Utilità e diletto. L'acqua e le residenze regie dell'Italia meridionale fra XI e XIII secolo." In L'eau dans la société médiévale: Fonctions, enjeux, images, edited by Jacques Dalarun. Rome, 1992. *Mélanges de l'École Française de Rome: Moyens âges, temps modernes* 104 (1992): 343–372.

Maroni, Alfredo. *Prime comunità cristiane e strade romane nei territori di Arezzo, Siena, Chiusi: dalle origini al secolo VIII.* Siena, 1973.

Mazzaoui, Maureen Fennell. " The Cotton Industry of Northern Italy in the Late Middle Ages: 1150-1450." *Journal of Economic History* 32 (1972): 262-286.

Menocal, María Rosa. "Pride and Prejudice in Medieval Studies: European and Oriental." *Hispanic Review* 53 (1985): 61–78.

———. *The Arabic Role in Medieval Literary History.* Philadelphia, 1987.

———. "And How 'Western' Was the Rest of Medieval Europe?" In *Américo Castro: The Impact of His Thought, Essays to Mark the Centenary of His Birth,* edited by Ronald E. Surtz, Jaime Ferrán, and Daniel P. Testa. Madison, 1988.

———. *Shards of Love: Exile and the Origins of the Lyric.* Durham, N.C., 1994.

Michelini, Tocci L. "Disegni e appunti autografi di Francesco di Giorgio in un codice del Taccola." In *Scritti di storia dell'arte in onore di Mario Salmi.* Vol. 2. Rome, 1962.

Ministerio dell'Interno, Pubblicazioni degli Archivi di Stato. *Archivio di Stato di Siena: Guida-Inventario dell'Archivio di Stato.* 4 vols. Rome, 1951–1995.

———. *Archivio del Consiglio generale del comune di Siena: Inventario.* Rome, 1952.

———. *Archivio del Concistoro del comune di Siena: Inventario.* Rome, 1952.

———. *Archivio della Biccherna del comune di Siena: Inventario.* Rome, 1953.

———. *Archivio di Balìa: Inventario.* Rome, 1957.

———. *Archivio notarile (1221–1862): Inventario.* Rome, 1975.

Minucci, Giotto. *La città di Grosseto e i suoi vescovi: già di Roselle A.D., 498-1988.* Florence, 1988.

Mitchell, Charles. "Ex Libris Kiriaci Anconitani." *Italia Mediovale e Umanistica* 5 (1962): 283–299.

Montreal Museum of Fine Arts. *Leonardo da Vinci: Engineer and Architect.* Montreal, 1987.

Moorman, John. *A History of the Franciscan Order from its Origins to the Year 1517.* Oxford, 1968.

Morandi, Ubaldo. *La Cattedrale di Siena. Ottavo centenario della consacrazione. 1179–1979.* Siena, 1979.

Morelli, Marcello, and Marco Tangheroni, eds. *Leonardo Fibonacci: Il tempo le opere, l'eredità scientifica.* (Pisa, 1994).

Moscadelli, Stefano. "Apparato burocratico e finanze del Comune di Siena sotto i Dodici (1355–1368)." *Bullettino senese di storia patria* 89 (1982): 29–118.

Muendel, John. "The 'French' Mill in Medieval Tuscany." *Journal of Medieval History* 10 (1984): 215-247

———. "Medieval Urban Renewal: The Communal Mills of the City of Florence, 1351–1382." *Journal of Urban History* 17 (1991): 363–389.

———. "The Internal Functions of a 14th-Century Florentine Flour Factory." *Technology and Culture* 32 (1991): 498-520.

Muller, Joseph. *Documenti sulle relazioni delle città toscane coll'Oriente cristiano e coi Turchi fino all'anno MDXXXI.* Florence, 1879.

Mumford, Lewis. *The City in History.* New York, 1961.

———. *Technics and Human Development.* Vol 1 of *The Myth of the Machine.* New York, 1967.

Musso, Gian Giacomo. *Navigazione e commercio genovese con il Levante nei documenti dell'Archivio di Stato di Genova (Secc. XIV-XV).* Rome, 1975.

Nante, N. "L'approvvigionamento idrico della città di Siena." *Rivista Italiana d'Igiene* 48 (1988): 175–198.

Nante, N., et al. "Tutela e gestione della risorsa idrica nella storia di Siena." *Rivista Italiana d'Igiene* 49 (1989): 59–74.

Needham, Joseph. *Introductory Orientations.* Vol. 1 of *Science and Civilization in China.* Cambridge, 1954.

Neuburger, Albert. *The Technical Arts and Sciences of the Ancients.* Translated by Henry L. Brose. London, 1930.

Occhini, Pier Ludovico. *Fabio Bargagli Petrucci: Podestà di Siena.* Florence, 1940.

Origo, Iris. "The Domestic Enemy: The Eastern Slaves in Tuscany in the Fourteenth and Fifteenth Centuries." *Speculum* 30 (1955): 321:366.

Origone, Sandra. "Mulini ad acqua in Liguria nel XV secolo." *Clio: rivista trimestrale di studi storici* 10 (1974): 89–120.

Owens, E. J. *The City in the Greek and Roman World.* London, 1991.

Paget, Robert F. *Central Italy: An Archaeological Guide: The Prehistoric, Villanovan, Etruscan, Samnite, Italic, and Roman Remains, and the Ancient Road Systems.* Park Ridge, N.J., 1973.

Painter, Sidney. "Western Europe on the Eve of the Crusades," in *The First Hundred Years,* ed. Marshall W. Baldwin, vol. 1 of *A History of the Crusades,* 2d ed., Kenneth M. Setton, ed. (Madison, 1969).

Palmer, Richard. "'In this Our Lightye and Learned Tyme': Italian Baths in the Era of the Renaissance." In *The Medical History of Waters and Spas,* edited by Roy Porter. Medical History Supplement, no. 10. London, 1990.

Pardi, G. "La popolazione di Siena e del territorio senese attraverso i secoli." *Bullettino senese di storia patria* 32 (1925): 3–62.

Parsons, William Barclay. *Engineers and Engineering in the Renaissance.* Baltimore, 1939.

Passeri, Vincenzo., ed. *Repertorio dei toponimi della Provincia di Siena, desunti dalla cartografia dell'Istituto Geografico Militare.* Siena 1983.

———. "Genesi e primo sviluppo del comune di Siena." *Bullettino senese di storia patria* 51–52 (1944–1947): 36–96.

Pepper, Simon, and Nicholas Adams. *Firearms and Fortifications: Military Architecture and Siege Warfare in Sixteenth-Century Siena.* Chicago, 1986.

Pertici, Petra. *La città magnificata: interventi edilizi a Siena nel Rinascimento: l'Ufficio dell'ornato (1428-1480).* Siena, 1995.

Pevsner, Nicholas. "The Term 'Architect' in the Middle Ages." *Speculum* 17 (1942): 549–562.

Phillips, Kyle M., Jr. *In the Hills of Tuscany: Recent Excavations at the Etruscan Site of Poggio Civitate (Murlo, Siena).* Philadelphia, 1993.

Piccinni, Gabriela. "L'ospedale di Santa Maria della Scala di Siena: Note sulle origini dell'assistenza sanitaria in Toscana (XIV-XV secolo)" in *Città e servizi sociali nell'Italia di secoli XII-XV. Atti del Convegno di Studio del Centro Italiano di studi di storia e d'arte di Pistoia, 1987.* Pistoia, 1990.

Pietramellara, Carla. *Il Duomo di Siena: Evoluzioine della forma dalle origini alla fine del trecento.* Florence, 1980.

Pingree, David. "The Diffusion of Arabic Magical Texts in Western Europe." In *La diffusione delle scienze islamiche nel Medio Evo europeo: (Roma, 2–4 ottobre 1984): convegno internazionale promosso dall'Accademia nazionale dei Lincei,* edited by Biancamaria Scarcia Amoretti. Rome, 1987.

Pinto, Giuliano. *La Toscana nel tardo medioevo: ambiente, economia rurale, società.* Florence, 1982.

———. "'Honour' and 'Profit': Landed Property and Trade in Medieval Siena." In *City and Countryside in Late Medieval and Renaissance Italy,* edited by Trevor Dean and Chris Wickham. London, 1988.

Plesner, Johan, with an introduction by Thomas Szabó. *Una rivoluzione stradale del dugento.* Florence, 1979.

Ploeg, Kees van der. *Art, Architecture, and Liturgy: Siena Cathedral in the Middle Ages.* Groningen, 1993.

Powers, James F. "Frontier Municipal Baths and Social Interaction in Thirteenth-Century Spain," *American Historical Review* 84 (1979): 649–667.

Prager, Frank D. "Fontana on Fountains: Venetian Hydraulics of 1418." *Physis: Rivista Internazionale di Storia della Scienza* 13 (1971): 341–360.

———. "A Manuscript of Taccola, Quoting Brunelleschi, on Problems of Inventors and Builders." *Proceedings of the American Philosophical Society* 112 (1968): 131–149.

Prager, Frank D., and Gustina Scaglia. *Brunelleschi: Studies of his Technology and Inventions.* Cambridge, Mass., 1970.

———. *Mariano Taccola and his Book "De ingeneis."* Cambridge, Mass., 1972.

"Precious Gallons," *Washington Post.* 21 August 1999, Final Edition, p. A 13.

Pryor, John H. *Geography, Technology, and War: Studies in the Maritime History of the Mediterranean, 649–1571.* Cambridge, 1988.

Quaroni, Ludovico. *Siena, centro storico: conservazione e recupero.* Milan, 1983

Redon, Odile. "Une forêt urbaine: la 'Selva del Lago.'" In *Villes, bonnes villes, cités et capitales: Études d'histoire urbaine (xii-xviii siècle) offertes à Bernard Chevalier,* edited by Monique Bourin. Tours, 1989.

———. "Le Contado de Sienne 1263–1270: Une frontière médiévale." *Mélanges d'Archéologie et d'Histoire de l'École Française de Rome* 87 (1975): 105–139.

———. *L'espace d'une cité: Sienne et le pays siennois (XIIIe-XIVe siècles).* Rome, 1994.

Renouard, Yves. "La Peste Noire de 1348–1350," *Revue de Paris* 57 (1950): 107–119.

Repetti, Emanuele. *Dizionario geografico fisico storico della Toscana. . . .* 6 vols. Florence, 1833–46.

Reti, Ladislao. "Francesco di Giorgio Martini's Treatise on Engineering and Its Plagiarists." *Technology and Culture* 4 (1963): 287–298.

———. *Tracce dei progetti perduti di Filippo Brunelleschi nel Codice Atlantico di Leonardo da Vinci.* Florence, 1964.

———. "Martini, Francesco di Giorgio." *Dictionary of Scientific Biography* 9:146–148.

Reynolds, Terry S. *Stronger than a Hundred Men: A History of the Vertical Water Wheel.* Baltimore, 1983.

Riccetti, Lucia. *La città costruita: lavori pubblici e immagine in Orvieto medievale.* Florence, 1992.

Riedl, Peter Anselm, Max Seidel, and Stefano Moscadelli, eds. *Die Kirchen von Siena.* 3 vols. to date. Munich, 1985-.

Rizzi, Alberto. *Vere da pozzo di Venezia: i puteali pubblici di Venezia e della sua laguna.* Venice, [1981].

Rizzitano, Umberto. "Gli Arabi in Italia." In *L'Occidente e L'Islam nell'Alto Medioevo,* edited by the Centro Italiano di studi sull'alto Medioevo. Spoleto, 1965.

Roaf, Susan. "Settlement Form and Qanat Routes in the Yazd Province," in *Qanat, Kariz, and Khattara: Traditional Water Systems in the Middle East and North Africa,* edited by Peter Beaumont, Michael Bonine, and Keith McLachlan (Wisbech, 1989).

Robins, F. W. *The Story of Water Supply.* London, 1946.

Rocchigiani, Roberto. "Urbanistica ed Igiene negli statuti Senesi del XIII e XIV secolo." *Studi Senesi* 70 (1958): 369–419.

Rogers, Everett. *Diffusion of Innovations.* 4th ed. New York, 1995.

Ron, Zvi Y. D. "Agricultural Terraces in the Judean Mountains." *Israel Exploration Journal* 16 (1966): 33–49, 111–122.

———. "Development and Management of Irrigation Systems in the Mountain Regions of the Holy Land." *Transactions of the Institute of British Geographers* 10 (1985): 149–169.

———. "Geohydrological Breakthrough in the First Jewish Temple Period and its Effect on the Development of Groundwater Resources." *Proceedings of the 12th Archaeological Conference in Israel.* Jerusalem, 1986.

———. "Qanats and Spring Flow Tunnels in the Holy Land." In *Qanat, Kariz, and Khattara: Traditional Water Systems in the Middle East and North Africa,* edited by Peter Beaumont, Michael Bonine, and Keith McLachlan. London, 1989.

Rose, Paul Lawrence. "The Taccola Manuscripts." *Physis* 10 (1968): 337–346.

Roselli, Piero, Brunella Ragoni, and Alfredo Forti. *Cartiere e opifici 'andanti ad acqua.* Florence, 1984.

Rosenbloom, Richard S. "Technological Innovation in Firms and Industries: An Assessment of the State of the Art." In, *Technological Innovation: A Critical Review of Current Knowledge,* edited by Patrick Kelly and Melvin Kranzberg. San Francisco, 1978.

Rubinstein, Nicolai. "Political Ideas in Sienese Art: The Frescoes by Ambrogio Lorenzetti and Taddeo di Bartolo in the Palazzo Pubblico." *Journal of the Warburg and Courtauld Institutes* 21 (1958): 179–207.

Rutigliano, Antonio. *Lorenzetti's Golden Mean: The Reformatori of Siena 1368–1385.* New York, 1991.

Saalman, Howard. *Medieval Cities: Medieval Architecture; European Architecture, 600-1200.* New York, 1962.

Sabine, Ernest L. "Butchering in Mediaeval London." *Speculum* 8 (1934): 335–353.

———. "Latrines and Cesspools in Mediaeval London." *Speculum* 9 (1934): 303–321.

———. "City Cleaning in Mediaeval London." *Speculum* 12 (1937): 19–43.

Sarton, George. *Appreciation of Ancient and Medieval Science During the Renaissance, 1450-1600.* Philadelphia, 1955.

———. *From Rabbi Ben Ezra to Roger Bacon.* Vol. 2 of *Introduction to the History of Science.* Baltimore, 1931. 491–495, 716–724.

Sauvaget, Jean. "Caravansérails syriens du moyen-âge." *Ars Islamica* 6 (1939): 48–55; 7 (1940): 1–19.

Scaglia, Gustina. *Francesco di Giorgio: Checklist and History of Manuscripts and Drawings in Autographs and Copies from ca. 1470 to 1687 and renewed copies (1764-1839).* Bethlehem, Penn., 1992.

———. "Autour de Francesco di Giorgio Martini, ingénieur et dessinateur." *Revue de l'art* 48 (1980): 7–25.

———. *Alle origini degli studi tecnologici di Leonardo.* Vinci, Italy, 1981.

———. "'Stanze-Stufe' e 'Stanze-Camini' nei 'Trattati' di Francesco di Giorgio di Siena." *Bolletino d'arte* 39–40 (1986): 161–184.

———. "Drawings of Machines for Architecture from Early Quattrocento in Italy." *Journal of the Society of Architectural Historians* 25 (1966): 90–114.

Scanlon, George T. "Housing and Sanitation: Some Aspects of Medieval Islamic Public Service." In *The Islamic City: A Colloquium,* edited by A. H. Hourani and S. M. Stern. Oxford, 1970.

Scanu, Giuseppe. ed. *Le Techniche dell'acqua. Atti del terzo convegno internazionale di studi geografico storici.* Núoro, 1990.

Schevill, Ferdinand. *History of Florence from the Founding of the City through the Renaissance.* New York, 1936. Reprint. New York, 1961.

———. *Siena: The History of a Mediaeval Italian Commune.* New York, 1909.

Schevill, Ferdinand, with an introduction by William Bowsky. *Siena: The History of a Mediaeval Italian Commune.* 2d ed. New York, 1964.

Schiaparelli, Attilio. *La casa fiorentina e i suoi arredi nei secoli XIV e XV.* Florence, 1908.

Schultz, Stanley K., and Clay McShane. "To Engineer the Metropolis: Sewers, Sanitation, and City Planning in Late-Nineteenth-Century America." *Journal of American History* 65 (1978): 389–411.

Seppilli, Anita. *Sacralità dell'acqua e sacrilegio dei ponti.* Palermo, 1985.

Serino, Vinicio. *Siena e l'acqua: storia e immagini di una città e delle sue fonti.* Siena, 1997.

Sestan, Ernesto. "Siena avanti Montaperti." *Bullettino senese di storia patria* 68 (1961): 25–74.

Settle, Thomas B. "The Tartaglia Ricci Problem: Towards a Study of the Technical Professional in the 16th Century." In *Cultura, scienze e tecniche nella Venezia del Cinquecento: Atti del convegno internazionale di studio Giovan Battista Benedetti e il suo tempo.* Venice, 1987.

Shelby, Lon R. "Mariano Taccola and His Books On Engines and Machines." *Technology and Culture* 16 (1975): 466-475.

Singer, Charles. "Epilogue: East and West in Retrospect." In *The Mediterranean Civilizations and the Middle Ages, c. 700 B.C. to c. 1500 A.D.* Vol. 2 of *A History of Technology.* Edited by Charles Singer, et al. Oxford, 1957.

Small, F. L. *The Influent and the Effluent: The History of Urban Water Supply and Sanitation.* Winnipeg, 1974.

Smith, Cyril Stanley. "Biringuccio, Vannoccio." *Dictionary of Scientific Biography* 2: 142–143.

Smith, Norman A. *Man and Water: A History of Hydro-technology.* New York, 1975.

Soricinelli, Paolo. *Storia sociale dell'acqua: riti e culture.* Milan, 1998.

Smith, Elizabeth Bradford, and Michael Wolfe, eds. *Technology and Resource Use in Medieval Europe: Cathedrals, Mills, and Mines.* Aldershot, 1997.

Squatriti, Paolo. "Water and Society in Late Antique and Early Medieval Italy A.D. 400–850." Ph.D. diss., University of Virginia, 1990.

———. *Water and Society in Early Medieval Italy, AD 400–1000.* Cambridge, 1998.

———. "'Advent and Conquests' of the Water Mill in Italy." In *Technology and Resource Use in Medieval Europe: Cathedrals, Mills, and Mines.* Edited by Elizabeth Bradford Smith and Michael Wolfe. Aldershot, 1997.

————, ed. *Working with Water in Medieval Europe: Technology and Resource-use* Leiden, 2000.

Squyres, Coy. H., ed. *Geology of Italy.* 2 vols. Tripoli, 1975.

Stallybrass, Peter. "Worn Worlds: Clothes, Mourning, and the Life of Things." *Yale Review* 81 (1993): 35–51.

Starn, Randolph. *Ambrogio Lorenzetti: The Palazzo Pubblico, Siena.* New York, 1994.

Sterpos, D. *Communicazioni stradali attraverso i tempi: Firenze-Roma.* Novara, 1964.

Stopani, Renato. *La via Francigena in Toscana: storia di una strada medievale.* Florence, 1984.

Strack, Henry. *Brick Architecture in Italy During the Renaissance and the Middle Ages with Original Sketches and Drawings.* Translated by Leigh Hunt. Berlin, 1889. Reprint. New York, 1910.

Stuard, Susan Mosher. "Ancillary Evidence for the Decline of Medieval Slavery." *Past and Present,* no. 149 (1995): 3–28.

Szabó, Thomas. "La rete stradale del contado di Siena: legislazione statutaria e amministrazione comunale nel Dugento." *MEFRM: MA* 87 (1975): 141–186.

Sznura, Frank. *L'espansione urbana di Firenze nel Dugento.* Florence, 1975.

Tangheroni, Marco. "Siena e il commercio internazionale nel duecento e nel trecento." In *Banchieri e mercanti di Siena,* edited by Franco Cardini. Rome, 1987.

Tarr, Joel A., and Gabriel Dupuy, eds. *Technology and the Rise of the Networked City in Europe and America.* Philadelphia, 1988.

Tarr, Joel A., and Josef W. Konvitz. "Patterns in the Development of the Urban Infrastructure." In *American Urbanism: A Historiographical Review,* edited by Howard Gillette, Jr., and Zane L. Miller. New York, 1987.

Thorndike, Lynn. "Sanitation, Baths, and Street Cleaning in the Middle Ages and Renaissance." *Speculum* 3 (1928): 192–203.

———. *A History of Magic and Experimental Science: Volumes III and IV Fourteenth and Fifteenth Centuries.* New York, 1934.

———. "Marianus Jacobus Taccola." *Archives internationales d'histoire des sciences* 8 (1955): 7–26.

———. "Giovanni da Fontana: The Application of Mechanical Ingenuity to Science," in *A History of Magic and Experimental Science: Volumes III and IV Fourteenth and Fifteenth Centuries.* New York, 1934.

Toledo (Spain). Ministerio de Obras Publicas e Ministerio de Educacion y Ciencia. *Estudio de la conservacion de los niveles del Tajo en Toledo.* Toledo, 1970.

Torriti, Piero. *Tutta Siena contrada per contrada: nuova guida illustrata storico-artistica della città e dintorni.* Siena, 1990.

Tortoli, Sandra. "Per la storia della produzione laniera a Siena nel Trecento e nei primi anni del Quattrocento." *Bullettino senese di storia patria* 82–83 (1975–1976): 220–238.

———. "I tiratoi dell'Arte della Lana di Siena nel Trecento: un contributo all'archeologia dell'industria manifatturiera." *Archeologia Medievale* 3 (1976): 413–428.

Tronzo, William, ed. *Intellectual Life at the Court of Frederick II Hohenstaufen.* Studies in the History of Art 44. Hanover, NH, 1994.

Tucci, Ugo. "Biringuccio, Vannoccio." in *Dizionario biografico italiano.* 10:625–631.

Turcan, Marie. "L'eau dans l'alimentation et la cuisine romaine." In *L'homme et l'eau en Méditérranée et au Proche Orient, III: L'eau dans les techniques séminaire de recherche 1981-1982,* edited by Pierre Louis. Travaux de la Maison de l'Orient, no. 11. Lyon, 1986.

Turrini, Patrizia. *Per honore et utile de la città: il comune e l'edilizia nel Quattrocento.* Siena, 1997.

Ubierna, José Antonio Juncà. "Tunnel Heritage in Spain: Roots of the Underground." *Tunnelling and Space Technology* 13 (1998): 131–141.

Udovitch, Abraham L. *Partnership and Profit in Medieval Islam.* Princeton, 1970.

Valle, A. della, ed. *I castelli del Senese: strutture fortificate dell'area senese-grossetana.* 2d ed. Siena, 1985.

Van Der Meer, Frédéric. *Atlas de L'Ordre Cistercien.* Amsterdam, 1965.

Vasari, Giorgio. *Le vite de' piu eccellenti pittori, scultori e architettori.* Edited by Aldo Rossi. Novara, 1967.

———. *Lives of the Painters, Sculptors, and Architects.* Translated by Gaston C. de Vere with and introduction and notes by David Ekeserdjian. 2d ed. 2 vols. New York, 1996.

Vendittelli, Marco. "Diritti ed impianti di pesca degli enti ecclesiastici romani tra X e XIII secolo." *L'eau dans la société médiévale: Fonctions, enjeux, images,* edited by Jacques Dalarun. Rome, 1992. *Mélanges de l'École Française de Rome: Moyens âges, temps modernes* 104 (1992): 387–430.

Vernet, Juan. "La Ciencia en el Islam y Occidente." In *L'Occidente e l'Islam nell'Alto Medioevo: 2 -8 April 1964,* edited by the Centro italiano di studi sull'alto Medioevo. Spoleto, 1965.

Vicchi, Pietro, ed. *Geo-Carte manoscritte e a stampa nell'Archivio di Stato di Siena: il fondo dei Quattro Conservatori.* Siena, 1990.

Viganò, Marino, ed. *Architetti e ingegneri militari italiani all'estero dal XV al XVIII secolo.* Castella 44. Livorno, 1994.

Vigarello, Georges. *Concepts of Cleanliness: Changing Attitudes in France since the Middle Ages.* Cambridge, 1988.

Waley, Daniel. *The Italian City-Republics.* 3d ed. London, 1988.

———. *Siena and the Sienese in the Thirteenth Century.* Cambridge, 1991.

Die Wasserversorgung antiker Stadte: Mensch und Wasser, Mitteleuropa, Thermen, Bau/Materialien, Hygiene. Vol. 3 of *Geschichte der Wasserversorgung.* Edited by the Frontinus Gesellschaft. Mainz am Rhein, 1988.

Ward-Perkins, J. B. *From Classical Antiquity to the Middle Ages: Urban Public Building and Central Italy A.D. 300-850.* New York, 1984.

———. *Landscape and History in Central Italy.* Oxford, 1964.

———. "Etruscan Engineering: Road Building, Water Supply, and Drainage." In *Hommages à Albert Grenier,* edited by Marcel Renard. Vol. 3 of *Collection Latomus* 58 (1962): 1636–43.

———. "Etruscan and Roman Roads in Southern Etruria." In *Journal of Roman Studies* 47 (1957): 139–43.

———. "Veii: The Historical Topography of the Ancient City." *Papers of the British School at Rome* 29 (1961): 47–52.

Waterer, John W. "Leather." In *The Mediterranean Civilizations and the Middle Ages, c. 700 B.C. to c. 1500 A.D.* Vol. 2 of *A History of Technology.* Edited by Charles Singer, et al. Oxford, 1957.

Watson, Andrew M. "The Arab Agricultural Revolution and Its Diffusion, 700–1100." *Journal of Economic History* 34 (1974): 8–35.

———. *Agricultural Innovation in the Early Islamic World: The Diffusion of Crops and Farming Techniques, 700-1100.* Cambridge, 1983.

Watt, William Montgomery. *The Influence of Islam on Medieval Europe.* Edinburgh, 1972.

Weller, Allen S. *Francesco di Giorgio, 1439–1501.* Chicago, 1943.

Welsh, Peter C. *Tanning in the United States to 1850: A Brief History.* Washington, 1964.

Wertime, Theodore A. "A Metallurgical Exploration through the Persian Desert." *Science* 159 (1968): 927–935.

White, Lynn jr. "Technology and Invention in the Middle Ages." *Speculum* 15 (1940): 141–159.

———. Review of *The Mediterranean Civilizations and the Middle Ages, c. 700 B.C. to c. 1500 A.D.,* edited by Charles Singer, et al., Oxford, 1957. In *Speculum* 33 (1958): 130–135.

———. "Tibet, India, Malaya as Sources of Western Medieval Technology." *American Historical Review* 65 (1960): 515–526.

———. *Medieval Technology and Social Change.* Oxford, 1962.

———. "Jacopo Aconcio as an Engineer." *American Historical Review* 72 (1967): 425–444.

———. "Medieval Borrowings from Further Asia." In *Medieval and Renaissance Studies: Proceedings of the Southeastern Institute of Medieval and Renaissance Studies, Summer 1969* [no. 5.]. Chapel Hill, 1971.

———. "Medieval Uses of Air." *Scientific American* 223 (1970): 92–100.

———. "Cultural Climates and Technological Advance." *Viator* 2 (1971): 171–191.

———. "The Flavor of Early Renaissance Technology." In *Developments in the Early Renaissance: Papers of the second annual conference of the Center of Medieval and Early Renaissance Studies. State University of New York at Binghamton, 4–5 May 1968,* edited by Bernard S. Levy. Albany, 1972.

———. "Indic Elements in the Iconography of Petrarch's *Trionfo della Morte.*" *Speculum* 49 (1975): 201–221.

———. "Medieval Engineering and the Sociology of Knowledge." *Pacific Historical Review* 44 (1975): 1-21.

———. "The Study of Medieval Technology, 1924–1974: Personal Reflections." *Technology and Culture* 16 (1975): 519–530.

———. "The Eurasian Context of Medieval Europe." In International Musiological Society. *Report of the Twelfth Congress Berkeley 1977.* London, 1981.

———. "Technology, Western." In *The Dictionary of the Middle Ages,* edited by Joseph L. Strayer. 13 vols. New York, 1982-1989. 12:650–664.

Whitney, Elspeth. *Paradise Restored: The Mechanical Arts from Antiquity through the Thirteenth Century.* Transactions of the American Philosophical Society, n.s., no. 80:1. Philadephia, 1990.

Wickham, C. J. *The Mountains and the City: The Tuscan Apennines in the Early Middle Ages.* Oxford, 1988.

Wieruszowski, Helene. "Art and the Commune in the Time of Dante." *Speculum* 19 (1944): 14–33.

Wikander, Örjan. *Exploitation of Water-power or Technological Stagnation?: A Reappraisal of the Productive Forces in the Roman Empire.* Lund, 1984.

Wulff, Hans. E. *The Traditional Crafts of Persia: Their Development, Technology, and Influence on Eastern and Western Civilizations.* Cambridge, Mass., 1966.

———. "The Qanats of Iran." *Scientific American* 118 (1968): 94–100, 105.

Wycherley, R. E. *How the Greeks Built Cities.* 2d ed. New York, 1969.

Zacchigna, Michele. *Sistemi d'acqua e mulini in Friuli fra i secoli XIV e XV: contributo alla storia dell'economia friulana nel bassomedioevo.* Venice, 1996.

Zanelli, Agostino. *Le schiave orientali a Firenze nei secoli XIV e XV.* Florence, 1885. Reprint. Bologna, 1976.

Zdekauer, Lodovico. *La vita privata dei senesi nel Dugento.* Siena, 1896.

————, ed. *Il mercante senese nel Dugento: Conferenza.* Siena, 1900.

Zupko, Ronald Edward, and Robert Anthony Laures. *Straws in the Wind: Medieval Urban Environmental Law—The Case of Northern Italy.* New York, 1996.

Index

B

D

N

O

P

Y

Z